Technological and Social Dimensions of the Green Revolution

Rising concerns about agricultural productivity and food security in rapidly changing economic and environmental contexts have led to renewed interest in agricultural development. But the extent to which new policies and programs will enable socially just and environmentally sustainable futures for rural communities remains a matter of intense debate. This book contributes to such debates by critically examining the intersection of agricultural histories, heterogeneous social contexts and new technological developments in rural communities across the Global South. It shows how experiences of the previous Green Revolution can inform new agricultural programs and enable equitable and participatory development in rural places. Through close engagement with rural communities, this book ensures that rural voices become part of the debate on agricultural development and suggests pathways for building on the gains of the Green Revolution without necessarily repeating its problematic social, technological and environmental aspects.

This book was published as a special issue of the *International Journal of Agricultural Sustainability*.

Pratyusha Basu conducts research on rural-urban intersections, international development and gender issues and she has published one book and numerous peer-reviewed journal articles on global to local aspects of contemporary rural transformations. Her research has been funded by the American Institute of Indian Studies and the U.S. Fulbright program.

Bruce A. Scholten researches the political economy of agriculture and dairy development in India and emerging economies, as well as organic farming in the US and UK. He is the author of two books and numerous peer-reviewed book chapters and journal articles.

Technological and Social Dimensions of the Green Revolution

Connecting Pasts and Futures

Edited by
Pratyusha Basu and Bruce A. Scholten

Routledge
Taylor & Francis Group
LONDON AND NEW YORK

First published 2013
by Routledge
2 Park Square, Milton Park, Abingdon, Oxon, OX14 4RN

Simultaneously published in the USA and Canada
by Routledge
711 Third Avenue, New York, NY 10017

Routledge is an imprint of the Taylor & Francis Group, an informa business

This book is a reproduction of the *International Journal of Agricultural Sustainability*, vol. 10, issue 2. The Publisher requests to those authors who may be citing this book to state, also, the bibliographical details of the special issue on which the book was based.

British Library Cataloguing in Publication Data
A catalogue record for this book is available from the British Library

ISBN13: 978-0-415-81518-5

Typeset in Times New Roman
by Taylor & Francis Books

Publisher's Note
The publisher would like to make readers aware that the chapters in this book may be referred to as articles as they are identical to the articles published in the special issue. The publisher accepts responsibility for any inconsistencies that may have arisen in the course of preparing this volume for print.

Contents

Citation Information

The chapters in this book were originally published in the *International Journal of Agricultural Sustainability*, volume 10, issue 2 (May 2012). When citing this material, please use the original page numbering for each article, as follows:

Chapter 1
Technological and social dimensions of the Green Revolution: connecting pasts and futures
Pratyusha Basu and Bruce A. Scholten
International Journal of Agricultural Sustainability, volume 10, issue 2 (May 2012) pp. 109-116

Chapter 2
Selling Guatemala's next Green Revolution: agricultural modernization and the politics of GM maize regulation
James Klepek
International Journal of Agricultural Sustainability, volume 10, issue 2 (May 2012) pp. 117-134

Chapter 3
Evaluating the Green Revolution after a decade: a Swaziland case study
Alan Terry
International Journal of Agricultural Sustainability, volume 10, issue 2 (May 2012) pp. 135-149

Chapter 4
Malawi's agricultural input subsidy: study of a Green Revolution-style strategy for food security
Marie Javdani
International Journal of Agricultural Sustainability, volume 10, issue 2 (May 2012) pp. 150-163

Chapter 5
Challenges for under-utilized crops illustrated by ricebean (Vigna umbellata) in India and Nepal
Peter Andersen
International Journal of Agricultural Sustainability, volume 10, issue 2 (May 2012) pp. 164-174

Chapter 6

Crop–livestock systems in rural development: linking India's Green and White Revolutions
Pratyusha Basu and Bruce A. Scholten
International Journal of Agricultural Sustainability, volume 10, issue 2 (May 2012) pp. 175-191

Chapter 7

Growing inequality: agricultural revolutions and the political ecology of rural development
Lucy Jarosz
International Journal of Agricultural Sustainability, volume 10, issue 2 (May 2012) pp. 192-199

Fig 1. Technology is appropriate to particular social, economic and environmental contexts. Above, tractors (photo by Bruce A. Scholten) such as this fuel efficient Massey Ferguson, on a California university organic farm, still function after five decades.

Fig 2. (photo by Pratyusha Basu) Crop-livestock systems provide these rice farmers in Gujarat, India with traction, fertiliser and food.

Introduction

Technological and social dimensions of the Green Revolution: connecting pasts and futures

Pratyusha Basu[a] and Bruce A. Scholten[b]

[a]*Department of Geography, Environment and Planning, University of South Florida, Tampa, FL, USA;* [b]*Science Laboratories, Department of Geography, Durham University, South Road, Durham, UK*

The 'Green Revolution', which has been marked by substantial and rapid increases in agricultural productivity from the 1950s onwards owing to high-yielding variety (HYV) seeds, chemical fertilizers and pesticides, large-scale irrigation and mechanization, as well as funding provided by international development aid agencies, is now embroiled in debates over possible pathways towards sustainable agricultural futures. Among the principal reasons for renewed debates is the decline in productivity gains since the 1990s as existing Green Revolution technologies reach their limits. Debates are also more broadly linked to the current financial crisis that has spurred the search for new spaces of accumulation and raised the value of rural resources (Harvey 2003, Moore 2008), and to food shortages that emerged due to the volatility of food prices, partly linked to the utilization of farmland for biofuels (McMichael 2010, Moseley *et al.* 2010).

This special issue provides a critical analysis of the technology–society nexus (borrowing from the science–society nexus in Jackson 2005) that characterized the Green Revolution across the Global South. This nexus is approached in terms of its construction, adoption and sustainability, which can also be considered to be the three stages in the evolution of the Green Revolution. The construction of the technology–society nexus is often based on unidirectional flows of knowledge from experts to farmers, and the institutional framework for the origin and spread of the Green Revolution consisted of an internationally coordinated alliance of scientists, governments, corporations and development agencies. Concerns over the control of knowledge have thus accompanied the Green Revolution, becoming more intense due to legally defined ownership of intellectual property in genetically modified organisms, forms of plant and animal species that are part of the new Gene Revolution (Buttel *et al.* 1985).

A focus on the adoption of Green Revolution technologies draws attention to the ways in which this was shaped by social inequalities. The continuing importance of land ownership, gender and indigenous identity, among other economic and social differences, in the ability to adopt and benefit from Green Revolution technologies, demonstrates that new technologies do not by themselves overturn entrenched economic and social orders (e.g. Chambers 2010). But the technology–society nexus is also transformative in terms of existing identities, so that new technologies are linked to new forms of political organization, ranging from agribusiness lobbies to peasant movements (e.g. Brass 1995), which in turn play a role in how Green Revolution technologies become unevenly distributed across national and international spaces.

A significant element in terms of renewing understandings of the technology–society nexus is the concept of agricultural sustainability (e.g. Schaller 1993), which extends the scientific paradigm of the Green Revolution through a participatory approach (Pretty 1999). Sustainable agricultural policies include a deliberate focus on marginalized small farmers, women and indigenous peoples, the incorporation of knowledge obtained from local experiences of agriculture, as well as support for working with natural processes rather than merely seeking to synthetically replicate them (Global Forum for Agricultural Research 2010).

The rest of this introduction connects the papers of this special issue to debates over the technology–society nexus by (i) situating the construction of the Green Revolution within institutional histories, including networks between governmental, non-governmental and corporate entities; (ii) highlighting the socially differentiated ways in which the Green Revolution was and is being adopted within local communities, especially in terms of differential access to technological inputs among small farmers; and (iii) utilizing existing knowledge of the Green Revolution towards building more sustainable agricultural programmes attentive to the economic, social and environmental needs of rural communities. The overall aim is not so much to provide generalized arguments regarding the Green Revolution's costs and benefits as much as to note that its costs are borne by particular social groups in particular localities. The geographical distribution of the costs of the Green Revolution therefore needs to be kept in mind, and this special issue seeks to examine the contextual characteristics of the Green Revolution through case studies drawn from across the Global South.

Constructing the Green and Gene Revolutions

The Green Revolution reflected some quintessential aspects of international development: first, it was organized around technology transfer so that development constituted a flow of knowledge from 'Western' scientists to the rural 'Third World'; second, such transfers were conducted under the aegis of the state, so that science and technology became a form of governmentality that enabled developing countries to represent themselves as modern. The prominence of Norman Borlaug, the American agricultural scientist who is often hailed as the 'Father of the Green Revolution' due to his role in first formulating HYV wheat and maize at Mexico's International Maize and Wheat Improvement Center (Centro Internacional de Mejoramiento de Maíz y Trigo), shows the key role played by scientists in constructing the Green Revolution. Similarly, within India M. S. Swaminathan is the agricultural scientist hailed as the 'Father of the Green Revolution', since his collaboration with Borlaug was instrumental in establishing HYVs in India. The figure of the agricultural scientist is thus the most visible representation of the Green Revolution. Although research was funded by the Rockefeller Foundation and Ford Foundation, Green Revolution technologies were disseminated by development agencies in partnership with governments, for instance through collaboration between the US, Mexico and India, so that the flow of scientific knowledge was encouraged by state support.

Underlying these scientific exchanges were Cold War imperatives (Escobar 1995, Perkins 1997), and the confluence between scientific knowledge and geopolitical power was a significant driver of the Green Revolution. The focus on South America and Asia in terms of agricultural development thus cannot be merely considered a natural stage of progression in the global dissemination of agricultural technologies. In fact, the Green Revolution was widely depicted as a means of containing communism, countering the political promises of socialism with the material bounties of capitalism. This political economy of the Green Revolution continues to shape current agricultural initiatives, as private and state capitalist investors seek to integrate remote areas into capitalist processes through accumulation by dispossession (Shiva 1991; Harvey 2003).

The latest round of technological innovation is genetic modification (GM), and one main institutional difference between earlier HYV and GM seeds is that the latter are more likely to be associated with corporate entities (Parayil 2003, Richards 2010). Thus, the US multinational, Monsanto, is one of the primary providers of GM seeds. Beginning as a supplier of agricultural chemicals, Monsanto's role in the production and control of genetic biotechnologies makes it a powerful entity shaping the future trajectories of agricultural growth (Parayil 2003). Current debates over GM seeds, and increasing use of related pesticides such as glyphosate (Binimelis *et al.* 2009), thus reflect concerns associated with the commercial harnessing of technology towards corporate profits and control, with such corporate control often only weakly regulated by the state. New government regulations therefore enable patents on biological life forms and processes, so that corporate control of knowledge is conducted in the realm of legality (Anderson and Tyers 1991, Lang and Heasman 2004).

James Klepek's paper in this special issue contrasts the Green and Gene Revolutions in Guatemala, revealing strategies deployed by Guatemala's agricultural companies in alliance with US governmental agencies to quell opposition to GM seeds. The significant issue here is that while the Green Revolution may have allowed some space for the continuation of indigenous agricultural practices, the Gene Revolution threatens to degrade the genetic inheritance of natural seeds, compromising the knowledge possessed by indigenous farmers as well as the spaces of resistance open to them. The deepening of market relations therefore is an especially significant aspect of the institutional framework of the Green Revolution (Dobbs 2008). Klepek usefully focuses on the ways in which pressures from above and below converge on Guatemala's government, further exacerbating the fragility of its democratic institutions. Although the power of Guatemala's private sector cannot be discounted, the role of the US in the construction of support for GM seeds points to continuities in geopolitical imperatives. As Klepek shows, Mayan indigenous groups oppose such state–corporate linkages, so that the burden of political protest often rests on the most marginalized segments in Guatemala, as it does across South America.

Adopting Green Revolution technologies: economic and social aspects

The Green Revolution ushered in an era of high productivity for staple food grains. Using the case of India, widely considered one of its success stories, yields for wheat increased from 750 kg/ha in 1953–1954 to 2,938 kg/ha in 2010–2011, whereas those for rice increased from 902 kg/ha in 1953–1954 to 2,240 kg/ha in 2010–2011 (Reserve Bank of India 2011). Such spectacular increases enabled India and other countries in Asia and Latin America to achieve food self-sufficiency and avert a Malthusian crisis, and the focus now is on bringing these benefits to Africa (Toenniessen *et al.* 2008, Bazuin *et al.* 2011). Not surprisingly, new GM seeds are also being promoted through a Malthusian discourse. Bill Gates, currently a powerful donor in international development, is reported to hold the position that 'countries can embrace modern seed technology and GM or their citizens will starve' (Blankinship 2012).

Abundance of production is only one aspect of the Green Revolution, and questions have been raised about the extent to which its benefits were available to all farmers. Thus, ethnographic studies in India document the ways in which relatively large farmers were more likely to adopt Green Revolution technologies, aggravating class and gender inequalities (Shiva 1988, Das 2002). The higher costs of Green Revolution inputs also led to political movements focused on the demands of richer farmers, including higher market support prices for their crops (Brass 1995, Gupta 1998). In the process, minimum wage rates for agricultural labour or the need for continuing land reforms were not on the agenda of these political formations.

Alan Terry and Marie Javdani draw on their community-level studies in Swaziland and Malawi, respectively, to emphasize the need to remain cognizant of the diverse social ramifications of new technologies. Their research extends the scope of Green Revolution analyses beyond the paradigmatic cases of South and Southeast Asia and South America. A greater understanding of the role of new agricultural technologies in Africa is especially compelling in the context of the formation in 2006 of an 'Alliance for a Green Revolution in Africa' (AGRA 2010), funded in large part by the Rockefeller Foundation and Bill and Melinda Gates Foundation (Toenniessen *et al.* 2008).

Terry provides a longitudinal analysis of the economic outcomes of the Green Revolution within a community transitioning from subsistence maize to irrigated sugarcane cultivation in Swaziland. His study shows the two sides of the narrative: smallholders who had access to new inputs through cooperative membership reported rising incomes across the 10 years of the study, while those without access due to lack of membership showed stagnant or declining incomes. Thus, even as the Green Revolution enabled some households to rise from poverty, its effects did not encompass the wider community, suggesting that the Green Revolution constructs new economic and social tensions as it unfolds. Terry's study also shows that a subtle analysis of the consequences of new technologies requires a long-term commitment to study sites.

Javdani's study further reveals the differences that emerge when attention is shifted from a generalized portrayal of production to the specifics of distribution in agricultural development. Malawi's programme of fertilizer subsidies has been lauded for increasing food production at the national level, but Javdani's household-level study shows that access to fertilizer is not sufficient to overcome land-based inequalities that structure local differences in food production and access. Similar to Terry therefore, Javdani's study shows that new technologies produce new forms of both wealth and inequalities, so that a new paradigm of agricultural development that can include both production and distribution within its purview needs to be formulated.

Taken together, Terry and Javdani argue that technological inputs reflect and often amplify existing social inequalities. This does not mean that agricultural production does not increase, or that such increases can be disregarded. Instead, these papers suggest that the unequal social distribution of the increases could constitute a limit to political support for new technologies. The new Green Revolution therefore should constitute a set of technological inputs within a specific policy framework that should seek to mitigate its socially differentiated outcomes. One prominent way in which solutions to the social dilemmas of the Green Revolution, and of development in general, are sought is through the notion of sustainability, and the final section of this introduction delves into this concept.

Towards sustainable agricultural environments

Sustainability in its classic definition is the process of ensuring that present-day needs do not compromise the ability of future generations to satisfy their needs (United Nations 1987). The value of this definition is that it demands a long-term perspective on economic development, a perspective often missing from capitalism's focus on short-term profits. However, to the extent that this definition contains a principle of moral responsibility that cannot be easily quantified, some misgivings have been expressed over its practical measurement (e.g. Schaller 1993, Hansen 1996). Lynam and Herdt (1989) have argued that measuring the sustainability of agricultural technologies is a different matter from measuring that of agricultural resources, since the former can be quantified in terms of environmental and social inputs and outputs, whereas the latter requires a consideration of community-level perspectives on proper resource distribution and use. Two approaches to sustainability can be distinguished: one that seeks to protect environmental values without completely denying the value of increasing production and another that argues that the formulation of sustainable policies requires a framework of social participation. These two approaches can be further examined through the notions of sustainable intensification and participatory development, both of which form underlying themes in this special issue.

Sustainable intensification is a process that seeks to balance productivity and environmental quality goals to prevent further encroachment by agriculture on forests and uncultivated spaces needed to mitigate anthropogenic impact on climate (Royal Society 2009, Pretty *et al.* 2011). Calls for an environment friendly Green Revolution have increased since the deleterious impacts of chemical-dependent agriculture became visible, ranging from chemical pollution of soils and water to impacts on human health (Singh 2000, Swaminathan 2000, Murgai *et al.* 2001, Tilman *et al.* 2002, Gupta 2007, Horlings and Marsden 2011, Sharma 2011). Whether such environmental costs are outweighed by the need to feed large populations remains a controversial issue, for instance in India where the majority of the population depends on agricultural livelihoods. Sustainable intensification is thus an attempt to encapsulate possibly competing objectives within a coherent framework. However, Rudel *et al.* (2009) have shown that increasing yields do not necessarily reduce the expansion of agricultural land, so that increases in farmland have proceeded despite technological increases in yields. They argue therefore that more deliberate policy interventions are required to ensure that technological gains are reflected in environmental benefits, such as the protection of forest land from agricultural growth.

Sustainable agriculture, however, is not only a matter of appropriate technologies, and the need to draw on the knowledge possessed by those most directly connected with farming is increasingly being recognized and emphasized (Hazell 2009). A practice-oriented definition of sustainable agriculture is provided by Pretty (1999, p. 259):

> '[the integration of] natural processes such as nutrient cycling, nitrogen fixation, soil regeneration and pest predators into food production processes, . . . [minimized] use of non-renewable inputs (pesticides and fertilisers) that damage the environment or harm the health of farmers and consumers . . . [and] better use of the knowledge and skills of farmers, so improving their self-reliance and capacities'.

Along the same lines, the Global Forum for Agricultural Research (2010), Royal Society (2009) and Foresight (2011) reports have called for greater involvement of hitherto marginalized rural social groups, especially small farmers and women, in the devising of sustainable agricultural systems, rather than have them merely serve as targets of sustainable technologies.

Peter Andersen's paper in this special issue provides an illuminating insight into the complex consequences of including local voices in agricultural decision making. Thus, even as loss of indigenous varieties is widely deplored as a deleterious consequence of the Green Revolution, Andersen shows that some communities in South Asia prefer new over native varieties of ricebean. This reveals the ability of local communities to adapt to the new, so that an attachment to traditional agriculture cannot be assumed *a priori*. As Lynam and Herdt (1989) argue, '[t]echnology modifies the sustainability of the system in which it is being applied; it is not an inherent characteristic of a variety, a cultural practice, or a particular input,' and this meshes with Foresight's (2011, p. 306) caution that '[i]t is counter-productive to be absolute or overly ideological about agricultural strategy'. Andersen's study thus suggests that if underutilized plant varieties, such as native ricebean species, are crucial to long-term ecological sustainability, then more research and development is needed to facilitate this outcome, rather than to assume that local communities will always support their traditional varieties and not respond to market signals.

Pratyusha Basu and Bruce Scholten's paper seeks to understand how the limits of the Green Revolution are already being addressed within rural communities in India through the combining of crops with livestock in local practices. Sustainable intensification here is thus expressed in mixed farming, so that the Green Revolution need not be isolated as a single solution to agricultural development. Rather, it constitutes one element in a multi-pronged approach for achieving income gains for smallholders and women. The point is that crop–livestock combinations found in rural communities across India do not reflect an overt linking of Green-White Revolutions in official policies, but emerge from community-level initiatives which draw on the strengths of various development programmes. Participation thus occurs through community institutions such as farmers' cooperatives, so that existing practices of farmers can be used to inspire changes in development programmes, rather than assuming that only formal participatory mechanisms can incorporate farmers' viewpoints. India's dairy programme also becomes important in the current development context since it is being promoted as a model for smallholders in other countries across Asia and Africa (Dugdill *et al.* forthcoming).

In conclusion, Lucy Jarosz situates this special issue within broader currents of contemporary agricultural development, from land grabs which threaten agricultural futures to food sovereignty movements that enable resistance. The Green Revolution and agriculture, in general, are now at a crossroads where all options remain open (e.g. Jayne *et al.* 2010, Viswanathan *et al.* 2012). Whether some options will be withdrawn in the future, in battles between the sustainability of corporate agriculture versus the sustainability of small farmers, cannot be easily predicted, although the balance of power leans heavily towards agribusiness corporations and their governmental allies. But, as the case studies of this special issue ultimately show, outcomes can surely be optimized by rethinking the Green Revolution and the practices swirling around its technology–society nexus.

Acknowledgments

This special issue first began as a series of papers presented at the 2009 Annual Meeting of the Association of American Geographers (AAG) in Las Vegas primarily in two sessions on the theme 'Following the Green Revolution: Ideologies, Inequalities, Offshoots'. The editors thank Zareen Bharucha for stellar editorial support, Sarah Pilgrim for first inviting us to build a special issue and Jules Pretty for encouragement throughout the process. Thanks are also due to Bill Moseley (for insights as a session discussant), Lucy Jarosz (for contributing a concluding paper), Peter Andersen and Marie Javdani (for papers from complementary sessions), Asmita Bhardwaj, Sharon Meagher, Amy Trauger, Sarah Halvorson and Ilona Moore (for presenting their papers at the original sessions), and last but not least, Alan Terry and James Klepek for being especially steadfast companions in the journey from special sessions to the special issue.

References

AGRA, 2010. *Alliance for a Green Revolution in Africa* [online]. Available from: http://www.agra-alliance.org/ [Accessed 27 Feb 2012].

Anderson, K. and Tyers, R., 1991. *Global effects of liberalising trade in farm products*. Trade Policy Research Centre, London: Harvester-Wheatsheaf.

Bazuin, S., Azadi, H., and Witlox, F., 2011. Application of GM crops in Sub-Saharan Africa: lessons learned from Green Revolution. *Biotechnology advances*, 29, 908–912.

Binimelis, R., Pengue, W., and Monterroso, I., 2009. 'Transgenic treadmill': responses to the emergence and spread of glyphosate-resistant Johnsongrass in Argentina. *Geoforum*, 40, 623–663.

Blankinship, D., 2012. *Gates defends focus on high-tech agriculture* [online]. Available from: http://articles.boston.com/2012-01-25/news/30663733_1_agriculture-genetic-modification-bill-gates [Accessed 27 Feb 2012].

Brass, T., ed., 1995. *New farmers' movements in India*. London: Frank Cass.

Buttel, F., Kenney, M., and Kloppenburg, J., 1985. From Green Revolution to biorevolution: some observations on the changing technological bases of economic transformation in the Third World. *Economic development and cultural change*, 34, 31–55.

Chambers, R., 2010. Beyond the Green Revolution: a selective essay. *In*: T. Bayliss-Smith and S. Wanmali, eds. *Understanding Green Revolutions: agrarian change and development planning in South Asia*. Cambridge: Cambridge University Press, 362–380.

Das, R., 2002. The Green Revolution and poverty: a theoretical and empirical examination of the relation between technology and society. *Geoforum*, 33, 55–72.
Dobbs, T.L., 2008. Commentary: what role for 'markets' in addressing the world food crisis? *International journal of agricultural sustainability*, 6 (3), 171–172.
Dugdill, B.T., Bennett, A., Phelan, J., and Scholten, B. Forthcoming. Dairy industry development programmes: their role in food and nutrition security and poverty reduction. *In:* E. Muehlhoff, A. Bennett and D. McMahon, eds. *Milk and dairy products in human nutrition*. Rome: FAO.
Escobar, A., 1995. *Encountering development: the making and unmaking of the Third World*. West Sussex: Princeton University Press.
Foresight, 2011. *The future of food and farming: final project report*. London: The Government Office for Science.
Global Forum for Agricultural Research, 2010. *Transforming agricultural research for development* [online]. Report for Global Conference on Agricultural Research 2010. Available from: http://gcardblog.files.wordpress.com/2010/03/gat-report-for-gcard-2010-version-11-0.pdf [Accessed 27 Feb 2012].
Gupta, A., 1998. *Postcolonial developments: agriculture in the making of modern India* [online]. Durham, NC: Duke University Press.
Gupta, S., 2007. *Ground water management in alluvial areas*. Available from: http://cgwb.gov.in/documents/papers/incidpapers/Paper%2011-%20sushil%20gupta.pdf [Accessed 27 Feb 2012].
Hansen, J., 1996. Is agricultural sustainability a useful concept? *Agricultural systems*, 50, 117–143.
Harvey, D., 2003. *The new imperialism*. Oxford: Oxford University Press.
Hazell, P., 2009. *The Asian Green Revolution* [online]. IFPRI Discussion Paper 00911. Available from: http://www.ifpri.org/sites/default/files/publications/ifpridp00911.pdf [Accessed 27 Feb 2012].
Horlings, I. and Marsden, T., 2011. Towards the real green revolution? Exploring the conceptual dimensions of a new ecological modernisation that could 'feed the world'. *Global environmental change*, 21, 441–452.
Jackson, S.A., 2005. The nexus: where science meets society. *Science* [online], 310, 1634–1639. Available from: http://www.sciencemag.org/content/310/5754/1634.full [Accessed 27 Feb 2012].
Jayne, T.S., Mather, D., and Mghenyi, E., 2010. Principal challenges confronting smallholder agriculture in Sub-Saharan Africa. *World development*, 38, 1384–1398.
Lang, T. and Heasman, M., 2004. *Food Wars: the global battle for mouths, minds and markets*. London: Earthscan.
Lynam, J. and Herdt, R., 1989. Sense and sustainability: sustainability as an objective in international agricultural research. *Agricultural economics*, 3, 381–398.
McMichael, P., 2010. Agrofuels in the food regime. *Journal of peasant studies*, 37 (4), 609–629.
Moore, J., 2008. Ecological crises and the agrarian question in world-historical perspective. *Monthly review*, 60, 54–62.
Moseley, W.G., Carney, J., and Becker, L., 2010. Neoliberal policy, rural livelihoods and urban food security in West Africa: a comparative study of The Gambia, Côte d'Ivoire and Mali. *Proceedings of the national academy of sciences*, 107, 5774–5779.
Murgai, R., Ali, M., and Byerlee, D., 2001. *Productivity growth and sustainability in post-Green Revolution agriculture in the Indian and Pakistan Punjabs*. Rome: FAO.
Parayil, G., 2003. Mapping technological trajectories of the Green Revolution and the Gene Revolution from modernization to globalization. *Research policy*, 32, 971–990.
Perkins, J., 1997. *Geopolitics and the Green Revolution: wheat, genes and the Cold War*. New York: Oxford University Press.
Pretty, J., 1999. Can sustainable agriculture feed Africa? New evidence on progress, processes and impacts. *Environment, development and sustainability*, 1, 253–274.
Pretty, J., Toulmin, C., and Williams, S., 2011. Sustainable intensification in African agriculture. *International journal of agricultural sustainability*, 9, 5–24.
Reserve Bank of India, 2011. *Handbook of statistics on Indian economy* [online]. Available from: http://rbi.org.in/scripts/PublicationsView.aspx?id=13609 [Accessed 27 Feb 2012].
Richards, D., 2010. Contradictions of the 'new Green Revolution': a view from South America's southern cone. *Globalizations*, 7, 563–576.
Royal Society, 2009. *Reaping the benefits: science and the sustainable intensification of global agriculture* [online]. Available from: http://royalsociety.org/uploadedFiles/Royal_Society_Content/policy/publications/2009/4294967719.pdf [Accessed 27 Feb 2012].
Rudel, T., *et al.*, 2009. Agricultural intensification and changes in cultivated areas, 1970–2005. *Proceedings of the national academy of sciences*, 106, 20675–20680.

Schaller, N., 1993. The concept of agricultural sustainability. *Agriculture, ecosystems and environment*, 46 (September), 89–97.

Sharma, V., 2011. Cancer Bathinda's dubious distinction. *The Hindu* [online]. Available from: http://www.thehindu.com/todays-paper/tp-national/article2427870.ece [Accessed 27 Feb 2012].

Shiva, V., 1988. *Staying alive: women, ecology and development in India*. Delhi: Kali for Women.

Shiva, V., 1991. *The violence of the Green Revolution: Third World agriculture, ecology and politics*. London: Zed, Penang, and Third World Network.

Singh, R., 2000. Environmental consequences of agricultural development: a case study from the Green Revolution state of Haryana, India. *Agriculture, ecosystems and environment*, 82, 97–103.

Swaminathan, M.S., 2000. An evergreen revolution. *Biologist*, 47, 85–89.

Tilman, D., *et al.*, 2002. Agricultural sustainability and intensive production practices. *Nature*, 418, 671–677.

Toenniessen, G., Adesina, A., and DeVries, J., 2008. Building an alliance for a Green Revolution in Africa. *Annals of the New York academy of sciences*, 1136, 233–242.

United Nations, 1987. *Report of the world commission on environment and development* [online]. Available from: http://www.un.org/documents/ga/res/42/ares42-187.htm [Accessed 27 Feb 2012].

Viswanathan, P., *et al.*, 2012. Agrarian transition and emerging challenges in Asian agriculture: a critical assessment. *Economic and political weekly*, 47, 41–50.

Selling Guatemala's next Green Revolution: agricultural modernization and the politics of GM maize regulation

James Klepek

Department of Environmental Studies, Rollins College, 1000 Holt Avenue, Campus Box 2753, Winter Park, FL 32789, USA

Although genetically modified (GM) agriculture is posited as key for increasing food production, controversy has limited the global adoption of the technology. There has been little work, however, analysing the negotiation of biotechnology regulations in developing countries where disputes are related to different histories of agricultural change from the US and Europe. Guatemala is a key case to examine the relationship of biotechnology debates to the historical trajectory of Green Revolution technologies. While Mayan farmers increased maize yields by mixing local seeds with modern fertilizers, the Green Revolution exacerbated structural inequities underlying Guatemala's agrarian system and heightened political tensions during the country's civil war. This paper contends that biotechnology regulatory politics are integrally tied to concerns over heightening existing tensions between small-scale and industrialized farming systems as well as to more recent issues of seed patenting. The first section links contemporary agrarian issues in Guatemala to agricultural modernization during the 1960s and 1970s. I then focus on international biotechnology disputes before turning to Guatemalan debates over GM maize. With continued struggles over biotechnology knowledge, I further suggest ways to envision an agricultural revolution more closely aligned with the goals of sustainability and the needs of small-scale farmers.

Introduction

Since 2007, Mayan indigenous and farmers movements have staged a series of protests in Guatemala calling on government support for maize producers in the wake of what Guatemalan president Álvaro Colom termed a 'national public calamity' of chronic food insecurity (Ochoa 2009). National and US business interests have promoted the adoption of genetically modified (GM) seeds to increase agricultural efficiency, address a national 'food crisis' and mitigate rural poverty. The emphasis on increasing domestic food production and encouraging rural development through new agricultural technologies is reminiscent of Guatemala's Green Revolution when farmers adopted commercial fertilizers and, to a lesser extent, hybrid maize seeds. Nonetheless, contemporary social movements have heavily resisted the commercialization of GM maize and state institutional disagreement reflects national debates over the technology. As of 2011, it is illegal to sell or grow GM maize in the country.

Resistance to agricultural biotechnology is not unique to the Guatemala case, with global controversy hindering the adoption of GM agriculture in many developing countries. Few studies, however, have attempted to understand how biotechnology debates are actually unfolding outside the US or Europe. Controversy in the US and Europe, where the vast majority of the population is not involved in agriculture, focuses on issues of consumer choice and the related issue of food labelling (Jasanoff 2005, Schurman and Munro 2010). By contrast, debates in Latin America, Asia and Africa centre on the threats that GM crops pose to food security and provision, as well as control over genetic resources and knowledge (Shiva 2001). Furthermore, disputes are related to different histories of agricultural change from industrialized countries and the continuing importance of small-scale farming systems to domestic food economies and subsistence survival (Alteri 2009).

Guatemala is a key case to examine the relationship of current biotechnology debates and controversy over GM maize, in particular, to the historical trajectory of the Green Revolution. Maize production remains of essential cultural and material importance to Mayan indigenous communities that make up 60% of the total population, and up to 90% in many rural areas of the Western highlands. Contemporary controversy is linked to the ways Mayan farmers negotiated agricultural modernization during the 1960s and 1970s as well as its limitations for the promotion of rural livelihoods. Peasant or *campesino* farmers increased maize yields through 'hybrid' practices of commercial fertilizer adoption, a continued reliance on traditional seeds and experimentation with modern seed varieties (Arias 1990, Van Etten and de Bruin 2007). Nonetheless, the Green Revolution also increased small-scale farmer vulnerabilities to global petrochemical market fluctuations and facilitated a broad expansion of plantation export agriculture, heightening historically rooted agrarian inequality and political tensions, which culminated in the most repressive period of the country's 36-year war (Manz 2004). With structural issues at the heart of the civil war unaddressed in the contemporary period, biotechnology debates centre on the extent to which the next Green Revolution offers a solution to or will further exacerbate rural inequality and food insecurity.

Guatemala further provides an important case to examine how the perceived role of the private sector in shaping contemporary agricultural biotechnologies as compared to the Green Revolution is shaping national biotechnology debates and state institutional politics. Among the most vocal proponents of the technology are the US-based biotechnology corporation Monsanto that acquired a Guatemalan maize seed company in 2008, and a Guatemalan sugar sector focused on creating biotech sugarcane varieties to spur ethanol expansion. Yet, while these sectors halted a United Nations-sponsored biosafety initiative that encouraged strict regulations to prevent potential risks of agricultural biotechnology to biodiversity and farmers, they have been unable to pass their own regulations, let alone commercialize GM crops.

As I illustrate in the following sections, understanding the past trajectory of the Green Revolution is essential to formulate a more critical perspective on contemporary agricultural biotechnologies. This paper contends that regulatory politics in developing countries and uncertainty over commercially growing GM crops are integrally tied to the continuities and changes of the 'Gene Revolution' from the Green Revolution. These include concerns over heightening existing tensions between small-scale and industrialized farming systems and the intensification of commercial input usage. They are also linked to new issues of patent enforcement and corporate control of the seed supply.

The next section focuses on the historical trajectory of the Green Revolution in Guatemala with particular emphasis on the gains and contradictions of this period for Mayan farmers. I then discuss the similarities and differences of the Green and Gene Revolutions that inform the general climate of international biotechnology disputes. Finally, I turn to Guatemalan controversy over GM agriculture and the politics of biotechnology regulation, with particular emphasis on conflict taking place within state institutions over the risks of biotech maize to biodiversity and rural livelihoods.

Research for this paper draws from interviews and participant observation carried out during 11 months of research in Guatemala in 2007 centred on biotechnology regulatory politics. For the purposes of this paper, I devote particular attention to interviews with government officials and the private sector organized under an Intersectoral Technical Commission directly involved in state discussions. Also, I draw from observations of key public and invitational events sponsored by the Technical Commission promoting the commercialization of GM crops as well by a coalition of indigenous and environmental movements organized under the National Network for Food Sovereignty (REDSAG) in opposition to GM maize. Finally, I draw extensively on historical literature as well as policy and technical documents to understand the international and Guatemalan contexts of the Green Revolution and biotechnology debates.

The Green and Gene Revolutions in context

While the dissemination of agricultural chemicals and hybrid seeds is generally credited as key to increasing global food supplies, critics argue that the Green Revolution heightened economic inequality and rural poverty (Shiva 1991, Yapa 1996), a trend that will intensify with agricultural biotechnology (Otero 2008). Related literature nonetheless points to the ways in which resource-based communities have utilized Green and Gene Revolution technologies to improve rural livelihoods (Bebbington 2000, Herring 2007). It is thus important to examine the historical and contemporary political and economic contexts in which particular technologies are deployed to understand their emancipatory and repressive potentials (Robbins 2004, p. 192). As I illustrate in Guatemala, the introduction of Green Revolution technologies resulted in both important gains and negative changes for small-scale agricultural producers that inform contemporary biotechnology debates.

Cold War era modernization and Guatemala's Green Revolution

The Guatemalan Green Revolution provides a particularly useful illustration of how existing political and economic contexts influence the promises and limitations of agricultural technologies. In 1954, a military coup reversed a decade of democratic governance in which the Guatemalan state attempted to redress long-standing inequality through land and labour reform. One of the first acts of the new dictatorship was the return of private and state land distributed in 1952 from the peasantry back to plantation owners while rural elites burned subsistence crops in their fields (Handy 1984, p. 187). In turn, the trajectory of successive post-1954 administrations pushed forward a vision of agricultural modernization that excluded progressive reform.

The role of the US Central Intelligence Agency in reinstating the military dictatorship and US policy to stem communist expansion via economic development made Guatemala one of the first test cases for Cold War era modernization in Latin America (Grandin 2004). From 1954 to 1957 alone, the US channelled \$100 million in economic aid and 'soft loans' for infrastructure projects and the diversification of the agro-export economy at a time when annual aid to the rest of the region was \$60 million (Handy 1984, p. 189). Building on a booming post World War II commodity market, the geographic expansion of export crops hinged on the broad adoption of fertilizers, pesticides and herbicides supplied by agro-chemical companies like Monsanto, which established Guatemalan subsidiary operations in 1965 (Tobis 1976, p. 276).

Challenging old and new plantation elite with land redistribution became politically infeasible so US development officials and national policy makers turned to new agricultural technologies to address food insecurity and political unrest. The Green Revolution solidified under the Alliance for Progress, created in 1961, in which the Kennedy administration pledged $20 billion in assistance to Latin America for political and economic modernization (Streeter 2006, p. 59).

In Guatemala, national use of chemical fertilizers increased nearly seven-fold during the 1960s and 1970s (FAOSTAT 2011). For Mayan indigenous communities, the application of new chemicals such as synthetic fertilizers and new hybrid maize seeds, combined with agricultural loans to small farmers, led to significant changes in maize production (Bulmer-Thomas 1987, p. 206). The United States Agency for International Development and local Catholic organizations disseminated fertilizers to Mayan farmers, leading to dramatic yield improvements in the Western highlands and marking a transformation of maize from solely a subsistence crop to a market commodity (Arias 1990, Carey 2009). At the national level, while land area dedicated to maize remained largely unchanged, production increased by 74% from 1952 to 1978 (Dunkerley 1988, p. 197).

Although the vast majority of Mayan maize producers in highland communities continued to rely on traditional seed saving and exchange, some interbred modern characteristics into farmer seeds. This was largely a response to increased fertilizer use, which resulted in taller, heavier maize plants susceptible to felling by strong winds. With little official emphasis on promoting hybrid seed usage, the dissemination of modern seeds occurred primarily at the informal level. Commercial hybrids developed in El Salvador and adopted throughout Central America were adapted by Guatemalan Mayans to highland growing conditions (Van Etten and de Bruin 2007, p. 61). The dissemination of modern seeds by the Institute of Agricultural Science and Technology, created in the early 1970s, was limited to lowland areas although they did support training in seed selection methods in some highland communities (Van Etten 2006, p. 703).

The adoption of Green Revolution technologies coincided with other factors that contributed to domestic maize production. The state opened land and initiated limited colonization projects in the agricultural frontier of the northern Petén region (Schwartz 1990). Rural cooperatives, with 27,000 members in 1967 (Manz 2004, p. 57), represented nascent political mobilization as peasants and an emerging rural Mayan middle class attempted to harness state-led modernization (Arias 1990). For the Guatemalan state, agricultural growth was a means to limit mobilization while preventing the expansion of a small, yet active guerilla insurgency during an escalating civil war.

Limitations of the Green Revolution

Despite these changes, the Green Revolution also exacerbated rural conflicts and heightened economic inequality. The increase in fertilizer consumption during the 1960s and 1970s most directly benefited the country's expanding plantation elite, evident with a rapid growth in large farms (Vilas 1995, p. 49). Land area dedicated to cotton on the southern coast increased from 5,000 hectares in 1948–1952 to 895,000 hectares in 1967 (Dunkerley 1988, p. 189) and sugarcane production rose 12 times, 1967–1976 (Vilas 1995, p. 43). Overall agro-export area increased 159%, 1948–1979 (Dunkerley 1988, pp. 187, 189). The contradictions of Guatemala's colonial legacy of land concentration became readily apparent by the late 1970s as agro-export expansion closed off possibilities to expand domestic maize production for a growing rural population. Although colonization provided opportunities for some of the landless peasantry, the geographic expansion of plantation elites largely resulted in the replication of an inequitable 'prevailing land tenure system' in the northern Petén region (Handy 1984, p. 213).

Communities that had become stakeholders in modernization further came up against a context of inequitable fiscal and credit policies as well as adverse external conditions (Bulmer-Thomas 1987, p. 207). Despite strong economic growth from 1960 to 1980, taxation and social spending in Guatemala were the lowest in Central America (Manz 2004, p. 49) and elites continuously blocked the fiscal reform necessary to support domestic food production (Bulmer-Thomas 1987, p. 184). The share of agricultural credit dedicated to coffee, cotton and sugar remained at 85% through the 1970s (Bulmer-Thomas 1987, p. 186). For most peasant producers, unofficial loans from labour contractors were the only means to purchase fertilizers to ensure subsistence survival on a shrinking land base (Manz 2004, p. 48).

With limited official support mechanisms and growing reliance on agro-chemicals, small-scale farmers were disproportionately impacted by a spike in fertilizer prices following the international oil crisis of 1973. In Guatemala, the cost of fertilizer rose from $50 per tonne in 1971 to $225 in 1973, increasing by another 215% in 1975 (Carey 2009, p. 304). This exacerbated economic polarization in indigenous communities and between the Mayan highlands and export-dominated coast. With fertilizer costs rising beyond local market maize prices, some became wage labourers or sharecroppers on land they previously owned while 'thousands upon thousands of campesinos were forced to migrate to the plantations and to the capital' (C'Amabal in Arias 1990, p. 241). For the remaining small-scale farmers, fertilizer use perpetuated, rather than resolved, a cycle of debt dependency (Carey 2009, p. 306).

In general, the increase in food production attributed to commercial fertilizers did help maintain food security with a national rising population as per capita maize consumption remained stable through the 1970s (FAOSTAT 2011). Yet, with 'medium- and large-scale farmers' responsible for over half of basic grain production at the end of the 1970s, 'it was clear that the increment in production has been achieved mainly at the expense of small-scale farmers whose share of total output declined' (Bulmer-Thomas 1987, pp. 206, 372). Furthermore, Guatemala experienced a strong rise in land inequality (Bulmer-Thomas 1987, p. 207) evident with a growth in *microfincas* (family farms not large enough to support subsistence survival) from 20.4% in 1964 to 41.1% in 1979 (Bulmer-Thomas 1987, p. 218). The doubling of Guatemalan microfincas represented the largest increase in Central America.

The Green Revolution exacerbated tensions between subsistence and industrialized production, but it also contributed to mobilization. Rural cooperatives increased from 145 groups in 1967 to 510 groups in 1976, with a membership totalling 132,000 people, over half located in the Mayan highlands (Dunkerley 1988, p. 474). Through the 1970s, rural cooperatives and political organization turned increasingly militant, eventually becoming a national political movement for land redistribution and labour rights (Arias 1990, Streeter 2006). In response, the military dictatorship initiated the most violent period of the country's 36-year civil war. From 1981 through 1983 alone, state-sponsored counterinsurgency practices killed 150,000 people, primarily rural Mayan civilians, completely destroyed hundreds of villages and displaced over a million people within Guatemala and neighbouring Mexico (CEH 1999, Jonas 2000). The destruction of maize fields and other subsistence crops was an essential component of the counterinsurgency against Mayan communities and one criterion utilized by the UN-sponsored Commission for Historical Clarification Committee to make its case for genocide committed by the Guatemalan state during the peace process. The destruction of indigenous communities and livelihoods resulted in a complete disarticulation of the Mayan political movement during the 1980s.

Guatemala in the early 21st century

The end of the civil war marked by the signing of Peace Accords in 1996 led to a re-emergence of indigenous activism around issues of cultural rights related to language and education (Hale 2002). Yet, structural problems at the heart of the historical conflict remain unaddressed. Guatemala currently has one of the highest degrees of economic inequality in Latin America, with 51% living below the poverty line and 15% in extreme poverty. Among the Mayan population, these numbers rise to 73 and 26%, respectively (PNUD 2010, p. 257). With half the population working in agriculture, estimates of national land concentration vary from 2% of people controlling 57% to 72% of total agricultural land (Gauster 2007, p. 1521).

Continued economic inequality, combined with free trade and market-based reform policies, has exacerbated a general if uneven historical trend of agro-export growth. A now powerful sugar sector tripled production and doubled land area dedicated to the crop on the Pacific Coast during the 1990s (Oglesby 2004, p. 558). Increased maize imports to Guatemala since the 1980s have offered an important 'escape valve' for US agricultural grain surpluses that has pushed down local prices and created a disincentive for domestic production (Van Etten and Fuentes 2004, p. 52). The government initiated extension services and credit for subsistence-based maize farmers in the mid-1980s as part of the development component of the counterinsurgency campaign that were nonetheless abandoned by the early 1990s.

Despite these issues, maize remains of fundamental cultural and material significance to rural communities, with small-scale farmers responsible for the majority of production. Contemporary Mayan practices of limited commercial fertilizer usage and the continued reliance on saved seeds in turn reflect the ways farmers appropriated as well as the limitations of Green Revolution technologies. By 2000 estimates, Guatemalan farmers utilized nearly half the fertilizer for maize production per capita compared to neighbouring Mexico (Aquino *et al.* 2000, p. 54). To the extent that small-scale maize farmers have engaged with the Green Revolution, they have done so in ways that intersect with rather than subsume traditional practice. Evidence indicates that rising costs and decreasing effectiveness have prompted Mayans to increasingly shift towards greater utilization of non-commercial, organic and locally available fertilizers (Carey 2009, p. 308).

In terms of the impacts of the Green Revolution on seed usage, estimates for the presence of farmer-saved maize seed as a total of national supply range from 91% (Aquino *et al.* 2000, p. 53) to 69% (Fuentes *et al.* 2005, p. 74), with commercial maize on the southern coast and northern Petén region increasingly relying on modern varieties. The area dedicated to hybrid maize or open-pollinated modern varieties does not exceed 1% in the Western highlands (Fuentes *et al.* 2005, p. 74). Yet, while small-scale farmers do not commonly buy seeds in the commercial market, modern seeds are not completely absent. In fact, the incorporation of modern seeds characteristics, such as shorter stalks and faster growing times, into local varieties by Mayan farmers is relatively common (Van Etten and de Bruin 2007).

Given these 'hybrid' practices of modern and traditional agriculture, it would seem logical to assume that agricultural biotechnology might follow a similar trajectory of limited yet significant adoption by peasant farmers. Nonetheless, GM agriculture has produced national debate over the technology related to broader issues of free trade and the future of subsistence farming. In particular, opposition to biotech maize has become a flashpoint for resistance to the US-backed Central American Free Trade Agreement ratified in 2005. Within this context of re-emerging social movements, market-based reform and rural inequality, national biotechnology debates centre on the extent to which GM maize will support or undermine food security and rural livelihoods. As I will show in the following sections, the similarities and differences of the Green Revolution from the proposed biotechnology revolution inform these debates over GM agriculture.

Global biotechnology debates

With commercialization since 1996, GM agriculture has become a multi-billion dollar industry. The total area of biotech crops has grown exponentially from 1.7 million hectares in 1996 to 148 million hectares in 2010 (James 2010). A highly consolidated biotech seed and chemical industry has been integral in shaping the trajectory of agricultural research and the commercial deployment of the technology particularly in the US, which is responsible for half of global biotech crop area (Boyd 2003, Kloppenburg 2005). Commercial production extends to 29 countries, with Latin America and Asia responsible for a growing share of GM crop area.

For some, agricultural biotechnology is key for the 'sustainable intensification' of agriculture necessary to increase food production in developing countries without significantly expanding land area (The Royal Society 2009). Proponents argue that currently available biotech varieties (insecticide-producing seeds and herbicide-tolerant GM varieties) reduce the need for expensive chemical inputs and act as a scale-neutral technology, which holds particular promise for small-scale farmers (Bailey 2005). Assessments of herbicide-tolerant soybeans in Argentina, for example, have found that 'the lower complexity' of production compared to conventional varieties (primarily through eliminating the need for weed tillage) has resulted in both productivity gains and environmental benefits (IAASTD 2009, p. 107). A number of studies recognize the potential and need for biotechnology innovation in developing countries to support small-scale farmers in tandem with less capital-intensive investment in agronomy and agroecology (Foresight 2011, p. 88). Research alliances between public institutions and transnational biotechnology corporations are viewed as essential to create new seed varieties adapted to adverse growing environments, including droughts caused by global climate change (Conway 2003).

Nonetheless, GM agriculture has also generated strong global controversy. Concerns over the health and environmental risks of GM crops have resulted in strict food labelling requirements, limited commercialization and mechanisms to segregate GMOs from organic and conventional agriculture in the European Union (Levidow and Boschert 2008). The following subsections focus on the major issues that inform unique emerging biotechnology debates in developing countries and that have relevance for Guatemala. These debates are in turn tied to the varied historical experiences of agricultural modernization during the Green Revolution as well as the continuing relevance of traditional agriculture for food production.

Agricultural industrialization, exports and bio-fuels

Like discussions of the Green Revolution, biotechnology debates in developing countries reflect disagreement over the implications of the technology for food security and rural livelihoods (McAfee 2008b, p. 148). A central point of concern is the extent to which GM agriculture will mitigate or accelerate commercial seed and agro-chemical reliance as well as heighten resource conflicts between large-scale and small-scale agricultural producers (Otero 2008). Some studies recognize both the potential of agricultural biotechnologies yet also assert that the development of the technology has favoured input-intensive industrialized production systems (IAASTD 2009, p. 92).

For critics, the disjuncture between the potential and actual trajectory of the technology is visible in Brazil and Argentina where large-scale farmers are responsible for roughly 60% of total biotech crop area in developing countries centred on herbicide-tolerant (round-up ready) soybeans for export to European and Chinese markets (Teubal 2008). Thus, the commercial deployment of GM crops is accelerating a 'technological treadmill' of commercial input reliance that disfavours small-scale farmers and is facilitating a new wave of industrial export expansion (Otero 2008, Binimelis *et al.* 2009). Current evidence suggests that biotech crops are exacerbating trends of rural inequality, land concentration and food insecurity in Latin America (Abramson 2009, Otero and Pechlaner 2009, p. 31) where small-scale farmers account for 41% of food production (rising to 51 and 77%, respectively, for staple crops of maize and bean) (Alteri 2009).

Biotechnology controversy is also linked to a new wave of bio-fuel or 'agro-fuel' expansion currently resulting in broad land-grabbing by transnational corporations in developing countries (Grain 2010). In particular, the creation of transgenic maize and sugarcane adapted to harsher climatic conditions is an important means to expand industrialized ethanol production in an era of accelerated climate change. Monsanto is officially releasing drought-tolerant maize varieties in the US in 2012 with plans for commercialization in Africa by 2014 (*The Economist* 2009). Although this next generation of GM crops is promoted as a means for increasing productivity and growing more food in developing countries, a concern is that the technology will contribute to this new wave of neo-colonial land concentration.

Environmental risks to agricultural diversity

Another source of controversy is the extent to which GM plant cross-pollination poses environmental risks to agricultural diversity in developing countries. This diversity reflects the dual importance of locally adapted seeds at the beginning (planting) and final stage of the farming cycle (consumption) with traditional seeds thus representing the 'alpha and omega' of subsistence livelihoods (Kloppenburg 2005, p. 37). Latin America is the largest global centre of agricultural diversity and where 40% of today's commercial food crops originate (McAfee 2008a, pp. 73–76). This includes maize, which is now the most widely produced crop in the world. Small-scale agricultural systems are responsible for maintaining the majority of maize diversity, the preservation of which is also essential for the continued improvement of commercial varieties (Soleri *et al.* 2006, p. 504).

Environmental risk debates are tied to perceptions of the similarities and/or differences in the molecular characteristics of biotech and Green Revolution seeds. Biotechnology proponents, adopting a US regulatory stance, assert that cross-pollination poses minimal risk to agricultural diversity in developing countries because biotech seeds are substantially equivalent to conventional varieties. Some ecologists nonetheless call attention to the potential environmental risks of biotech crops because gene flow from modern seeds has in some cases negatively impacted traditional agricultural systems and GM agriculture holds distinct environmental risks to biodiversity compared to conventional varieties (Soleri *et al.* 2006, p. 510).

In the first respect, the transfer of biotech traits via pollen through a process of sexual or 'vertical' gene flow, identical to processes that take place in conventional plants, has generated controversy over the potential of GM crops to degrade and out-compete local plants. As described by Alteri, the contamination of centres of agricultural diversity could result in the loss or replacement of local seed adaptations like 'resistance to drought, competitive ability, performance in polycrop systems, storage quality, etc' that 'produce relatively stable yields with a minimum of external inputs under changing environments' with less useful and potentially dangerous transgenic traits intended for industrialized agricultural systems (Alteri 2009, p. 1).

One unique potential environmental risk of GM agriculture compared to conventional Green Revolution seeds is 'horizontal' gene flow. This form of trait transfer involves the asexual transmission of transgenic material to other non-target plant or animal organisms (Giovannetti 2003) with potential to 'form novel, unintended and possibly dangerous combinations' (McAfee 2003, p. 208). A study illustrating the absence of herbicide-resistant genes in honey-bee populations emphasized the improbability that this type of genetic transfer could occur (Mohr and Tebbe 2007). Nonetheless, the potential for horizontal gene transfer has been posed as a significant issue for consideration in future environmental risk assessment (Chandler and Dunwell 2008, p. 28).

Critics have also raised questions on the suitability of US risk assessment models, which 'underestimate or ignore the dispersal of transgenic artifacts by natural means' (McAfee 2008a, p. 78). For example, a study by the Union of Concerned Scientists found significant transgenic contamination of non-GM maize varieties in the US, arguing it reflected a 'symptom of generally porous seed production and distribution systems' (Mellon and Rissler 2004, p. 2). The case of Mexico, in particular, has raised broader questions over the differentiated distribution of risks of GM agriculture to biodiversity in Latin America compared to the US and Europe (McAfee

2008b, p. 74). A 2001 study found that illegally planted GM maize, imported under free trade, had cross-pollinated with local varieties in a country that is a centre of origin for the species (Quist and Chapela 2001). The confirmation of the study's findings of continued transgene presence in maize landraces (Piñero-Nelson *et al.* 2009) illustrates both the incipient development of scientific methodologies for measuring GM maize cross-pollination in small-scale agricultural systems and the difficulty of removing biotech traits from landraces in small-scale agricultural systems characterized by high natural gene flow (Mercer and Wainwright 2008).

Seed patents

Developing country debates are further informed by new concerns over the corporate control of seed supplies through the extension of legal patent enforcement mechanisms. The continuing importance of traditional seeds in developing countries reflects advantageous characteristics compared to modern varieties, including lower yet more stable yields, less reliance on external inputs like agro-chemicals and irrigation, and a high adaptation to diverse and often marginal growing environments (Soleri *et al.* 2005, p. 142). Furthermore, the distinction between 'commercial' and 'traditional' seed varieties is not absolute as farmers in developing countries have incorporated desired hybrid seed traits into traditional varieties that are, in turn, selected, saved and replanted (Soleri *et al.* 2006, p. 505). Indeed, it is precisely the simultaneous resistance to and appropriation of Green Revolution seeds that underlies concerns over contemporary agricultural biotechnologies.

At a broader level, concerns over patenting speak to a key distinction between the Green Revolution largely initiated by public institutions and a Gene Revolution that, notwithstanding efforts by organizations like the Bill and Melinda Gates Foundation, is largely corporate led. Today's 'life science' corporations, which began as chemical companies and acquired seed companies during the 1990s (Boyd 2003), are responsible for 85% of global investment in agricultural biotechnology (IAASTD 2009, p. 30). Monsanto, now the dominant global leader in the GM seed market, owned 188 patents on maize seed and 266 patents on soybean in 2001 alone (Boyd 2003, p. 45). By 2011, 'round-up ready' transgenic seeds resistant to Monsanto herbicide accounted for the majority of the company's $11.8 billion in annual sales (*The Economist* 2009, Monsanto 2011, p. 4). In the US and Canada, the company has established ideal 'market conditions' through the rigorous legal prosecution of farmers who violate technology agreements by replanting GM seeds (Barlett and Steele 2008). Estimates that Monsanto seeds account for 88% of the total biotech crop planted area and 97% of the market share for GM maize in the US has spurred an ongoing anti-trust investigation (*The Economist* 2009, Otero and Pechlaner 2009).

The contemporary and potential future ability of corporations to enforce seed patents has become an important issue in debates over the impacts of the technology in developing countries (McAfee 1999, Herring 2007). In Mesoamerica, critics argue that the unregulated contamination of local maize with GM pollen, combined with the extension and enforcement of seed proprietary rights through the proliferation of US free trade agreements, represents a threat to the survival of small-scale agricultural systems (Desantis 2003). The dual means in which traditional seeds could come to contain GM traits (unintended cross-pollination or intentional farmer experimentation and appropriation) further speak to the interlinked environmental and socio-economic issues raised by contemporary agricultural biotechnologies.

The politics of agricultural biotechnology in Guatemala

While literature from the previous section is essential for understanding the real and potential future impacts of GM agriculture, it provides less insight into the actual negotiation of regulations in developing countries. The following discussion brings together existing work on debates over biotechnology's implications for Third World agricultural systems with emerging literature on the multi-faceted political struggles taking place over biotechnology science in state institutions and public arenas (Essex 2008, Jansen and Roquas 2008, McAfee 2008b, Fitting 2011). Instead of assuming regulation as simply a reflection of competing international interests, whether it be US promotion or European precaution, it is essential to recognize the wide range of actors involved in political struggles over biotechnology knowledge, including state scientists, transnational seed companies and indigenous movements.

I argue that agricultural biotechnology regulation is an interesting example to examine what McCarthy and Prudham (2004) view as the under-explored relationship between free market policy and environmental politics precisely because regulation is linked to the ability of various interests to put forward 'scientific' data when scientific knowledge itself is contested. While emphasizing commercial efforts to promote GM agriculture in Guatemala, I also examine factors that explain continued controversy over the technology including institutional conflict and social-movement opposition.

Corporate interests and the Technical Commission

Guatemala has the most strategic private sector interests in Central America promoting the commercialization of GM agriculture. This includes the largest seed company in the Latin American tropics, *Semillas Cristiani-Burkard* (SCB), headquartered in Guatemala. Founded in El Salvador in 1966, the company transferred to Guatemala in 1980 and soon extended markets for hybrid maize to 12 countries in Central America, the Caribbean and South America in partnership with seed and agro-chemical giants Pioneer-HiBred and Monsanto. In 2007, the company was utilizing technology from Monsanto to develop transgenic maize varieties with plans to release them for commercial sale by 2012 (Klepek 2007). The acquisition of the company by Monsanto in 2008 marked a solidification of US agricultural biotechnology interests in Mesoamerica, with the company poised to control an emerging regional market for GM seeds. With food security once again a priority since the 2007 food crisis, SCB believes that the legalization of transgenic seeds in Guatemala will trigger the industrialization of maize production throughout the region.

Another group actively promoting GM agriculture is the Guatemalan sugar sector, concentrated in the hands of 14 families who control much of the 215,000 hectares dedicated to sugarcane production. The economic significance of sugar, and as of 2005 sugar-based ethanol, is particularly evident in Guatemala compared to other countries in Central America. Guatemala controls eight of the top. 13 sugar refineries, produces half of the region's sugar, and accounts for 44% of Central America's ethanol. Guatemala is also the fifth biggest exporter of sugar in the world, with a dramatic increase in sugar sales and production driven by growing demand for ethanol in US and European markets in the 2000s (Tay 2009, p. 4). The sector is currently investigating transgenic cane varieties to facilitate expansion into new climatic regions to meet growing international demand for ethanol. Estimates see the potential to extend the 215,000 hectares currently planted by between 135,000 and 600,000 hectares (Fradejas *et al.* 2008, p. 29; Tay 2009, p. 3). Promoted by US trade representatives as 'an invaluable opportunity' for crop diversification, rural employment and poverty reduction (Tay 2009, p. 7), critics argue that current biofuel expansion is replicating the highly consolidated and inequitable plantation system on the southern coast (Solano 2008).

In 2005, the Guatemalan private sector in tandem with the US Department of Agriculture's Foreign Agricultural Service (USDA-FAS) converged to effectively lobby state agencies to stop a UN-sponsored biosafety legislative proposal for its emphasis on the potential environmental and socio-economic risks of agricultural biotechnology. Of particular concern was that the biosafety proposal posed a 'series of contradictions' with US trade obligations for proposing a regulatory framework in which Guatemalan decision makers, with input from oppositional social movements, could refuse imports or commercial planting regardless of approval in the US (Huete 2005). In 2007 and 2008, these interests began channelling efforts to promote agricultural biotechnology in public discussions and state institutional regulations through an Intersectoral Technical Commission on Biotechnology. The private sector, most prominent of which were representatives from SCB, the sugar sector and the food industry, comprised 10 of the commission's 20 members and controlled the three-person governing committee, with representatives from government and academia making up the other half.

Public events and news conferences sponsored by the Technical Commission largely promoted agricultural biotechnology to urban-based students, policy makers, commercial groups and scientific experts. In rural areas, the National Seed Fair was significant as the first major event attempting to garner support for GM seeds among rural farmers. The three-day conference was held in April 2007 on the southern coast of Guatemala in Retalthuleu, an area dominated by agro-export production but also the site of some industrialized commercial maize farming. I attended the event through an invitation from SCB. Besides SCB, the Fair included the participation of the state Ministry of Livestock and Agriculture (MAGA) and major international agrochemical and biotechnology firms Monsanto, Syngenta, Bayer and Dupont.

While most of the public presentations addressed the proper use of agro-chemicals, hybrid seeds and heavy machinery, the director of SCB focused on debunking the environmental 'myths' of GM maize and promoting the economic benefits of the technology. In a presentation entitled 'Biotechnology in Tropical Maize Varieties', Cristiani deemphasized scientific concerns over introducing GM maize into the environment by comparing the spread of GM traits to domestic varieties as similar to the exchange of genes in traditional farmer seed selection, arguing that new biotech genes would enhance biodiversity (Cristiani 2007). While recognizing the importance of preserving agro-diversity, the director nonetheless assumed an inevitable displacement of traditional or *criolla* seeds through industrialization. According to Cristiani:

> It is necessary to have adequate seed banks to be able to conserve our biodiversity. In the long term, it is not sustainable to continue growing criollas forever. If producers wish to continue forward and have higher productivity to be profitable, they cannot keep growing criollas (Cristiani 2007).

If predictions on the widespread adoption of GM agriculture were correct, germplasm banks would preserve hybrids and local seeds necessary for public research and development while facilitating greater private control over the seed supply, potentially increasing the divide between 'traditional' and large-scale maize production. As the primary spokesperson for national and transnational seed interests in Guatemala, SCB's promotion of GM agriculture thus hinges on portraying cross-pollination, as well as the marginalization of local maize varieties and subsistence practices, as a natural, if not desirable, component of agricultural modernization.

Disagreements over science-based regulatory policy

In response to the UN legislation and popular opposition, the Technical Commission further engaged with government ministries through private meetings and invitational conferences. In July 2007, the Technical Commission facilitated a Symposium on Science-Based Regulatory Policy focused on 'developing a model for risk analysis' for agricultural biosafety (SENACYT 2007). The event was sponsored by the USDA-FAS, led by three US academics and attended by representatives from government ministries involved in biotechnology regulation as well as the private sector. Along with a local radio host, I was the only investigator permitted to participate in the entire duration of the symposium, which included a series of presentations followed by a roundtable discussion.

While the formal title suggested the symposium would focus on facilitating 'science-based regulation', there was little discussion on 'science' per se, such as drafting environmental risk assessment procedures in the Guatemalan context, for example. Taking the expertise of US regulatory assertions as given, the event was more accurately a political event intended to convince government agencies to facilitate the full commercialization of GM plants.

The language of the commentaries during the closing meeting nonetheless reflected competing views within the Technical Commission over how best to realize these goals. According to one official:

> Laws must be a product of demand and are created when there is something already existing. It would be better to go to the respective agencies, make sure particular regulations are in place and then work with investigators and seed producers to begin growing transgenics (SENACYT 2007).

In contrast, another member expressed concerns stating that 'a ministerial accord is a good place to start for investigation and production but the question is how to gain more legal certainty without having to go through Congress and the whole political environment' (SENACYT 2007). With continuing national controversy over the technology and uncertainty over the future of the UN programme, establishing a pro-commercial Guatemalan law was seen as important yet politically difficult.

Though most of the discussion centred on creating a system of regulations through interagency negotiation, one food industry representative from the Technical Commission suggested planting biotech crops immediately, and then returning to regulatory negotiation.

> In my view, it is easier to ask for forgiveness than permission. We need to develop the technology through the private sector and in the next two to three months, create a cooperative growing GM maize. Then the UN programme will not be relevant (SENACYT 2007).

Stretching the boundaries of legality was thus seen as a possible means to naturalize what others portray as potentially dangerous genetic contamination. In particular, commercial interests proposed utilizing the unique characteristics of maize itself (the plant's ability to cross-pollinate making it difficult to control) to shape regulations and facilitate the commercialization of GM agriculture. The pro-active attitude of some private sector members of the Technical Commission speaks to the politics of institutional regulatory discussions and the role of trade and agro-capital interests in attempts to articulate the scientific discourse of state ministries away from more precautionary alternatives.

Yet, division over the regulation of GM agriculture extended both between state agencies and perhaps most significantly within the Technical Commission itself. I encountered evident concern from commission representatives over the environmental risks of GM maize. One of these included a molecular biologist on the Technical Commission who had diverse professional experience in both academia and the private sector. Emphasizing the potential benefits of GM agriculture in reducing chemical usage and allowing greater productivity, the representative nonetheless asserted:

> How is the US supposed to make recommendations in a country that has a complete lack of infrastructure to manage the risks of agricultural biotechnology?. . .The economic interests driving the biotechnology agenda and promoting a US-style regulatory system do not have the best interests of Guatemala in mind (Guatemalan biotechnology consultant, personal communication number 23, Guatemala City, 25 June 2007).

As further acknowledged by one USDA-FAS expert, future regulations may need to consider Guatemala's 'unique environmental sensitivities, particularly with respect to diversity of maize germplasm' (Kardwell 2005, p. 3).

Tensions within the Technical Commission included concerns over the potential environmental risks of GM maize to Guatemalan biodiversity as well as the role of the private sector in shaping regulations and the future trajectory of biotechnology investigation and commercialization. As a US-trained biotechnology scientist and leading member of the Technical Commission asserted:

> Private sector representatives from big industry are mainly interested in having little regulation. . .I am not expressing frustration towards Monsanto but I believe it should be the government's role to develop seed varieties like this. The issue I struggle most with is biodiversity and cross-pollination. Is it responsible to make transgenics of food that Guatemalans grow? (National Council of Science and Technology (CONCYT), personal communication number 20, Guatemala City, 27 April 2007).

This question reflects concerns over protecting local biodiversity, the role of public versus private development of the technology and the extent to which GM maize would help alleviate or exacerbate tensions between large-scale and subsistence-oriented agriculture. The answers to these questions, though portrayed in public spaces as determined by agreed-upon 'scientific' facts and norms, were fraught with uncertainty over environmental risks and diverging political economic perspectives within state institutions and the Technical Commission.

Popular mobilization

Uncertainty over the future of agricultural biotechnology is not simply reflective of 'expert' disagreement within state institutions, but also growing social-movement opposition. The most politically organized and vocal opponent of GM maize is the REDSAG, a coalition of 200 nationally based Mayan, peasant and environmental groups mobilized around issues related to small-scale agriculture. An important aspect of REDSAG's political campaign has been to host a series of 'native seed fairs' focused on mobilizing rural communities against GM agriculture and strengthening local agricultural production through the exchange of traditional seeds and knowledge.

In April 2007, I attended the *Feria Nacional de Semillas Nativas* or the 'National Native Seed Fair' held in the town of Solola. The fair included participation of thousands of Mayan farmers and community leaders and was attended by local and national news media sources. It was the largest out of a series of meetings in 2007 and 2008 focused on rural mobilization in opposition to GM maize and preserving native seeds. The fair offered diverse forums, ranging from presentations to seed exchanges and visits to local farms, for observing and participating in popular efforts to influence biotechnology debates and wider discussions over free trade and domestic agricultural production.

Figure 1, an illustration displayed at the 2007 seed fair entitled 'Struggle for Food Sovereignty', is particularly emblematic of broader themes evident in social-movement opposition to the technology. On one side, it depicts a traditional Mayan surrounded by the four-colour varieties of Guatemalan maize with the phrases: 'We conserve what we inherited from our ancestors; We protect our Mother Earth'. The other side of the illustration tells of the perceived dangers of GM crops to this heritage of Mayan maize production. The phrase 'Caution! Transgenics are a threat to our biodiversity' is placed next to an ear of genetically engineered maize and a US corporate giant enchaining a Mayan peasant. Of equal importance is the evident linkage between biotech crops and agro-chemical dependency portrayed as poisoning the soil. In essence, the perceived threats posed by agricultural biotechnology, connected to farmer concerns over intensification of the

Figure 1. REDSAG banner, native seed fair, 2007.
Source: Photo by author.

Green Revolution and new corporate control of the seed supply, are interlinked in contemporary public debates over the technology.

Concluding remarks

This paper has argued that understanding the historical trajectory of the Green Revolution is important for developing a critical perspective on the possibilities and contradictions of GM crops. While emphasizing how the Green Revolution facilitated the expansion of elite-orientated agriculture, my aim has also been to illustrate the ways Mayans harnessed agricultural modernization, as well as the limitations of these technologies, in promoting rural livelihoods. Small-scale farmers utilized fertilizers to increase maize production, yet were disproportionately hurt by rising fertilizer costs. They incorporated traits from modern seeds while continuing to rely on locally developed varieties and remaining outside the formal seed market.

This history is reflected in the contemporary context of Mayan maize production characterized by limited utilization of commercial fertilizers and the continued importance of seed saving and exchange – practices which holders of GM patents would like to eliminate through the realization of a particular vision for agricultural modernization. Trade and commercial interests have attempted to quell debates by naturalizing what critics view as genetic contamination and the inevitable disappearance of traditional seeds. This context, in turn, informs institutional and popular concerns over the potential risks of GM maize to biodiversity and subsistence livelihoods.

Beyond these specificities, the Guatemalan controversy provides an important window to understand the varied resistance and/or appropriation of GM agriculture in the Third World as integrally linked to the similarities and differences of the technology from the historical Green Revolution. The issues of seed patenting, the corporate control of the technology and the exacerbation of the technological treadmill of industrialized agriculture are common concerns evident in Guatemala and manifested throughout the Third World. These issues are also important for understanding how agricultural biotechnology could potentially be disentangled from transnational commercial interests to better benefit small-scale agricultural systems reliant on seed saving. For example, Herring (2007) provides a poignant example of how farmers in India and Brazil (the former a significant producer of Bt cotton and the latter of herbicide-tolerant soybean) are circumventing international seed proprietary obligations and national biosafety regulations through locally based commercial seed pirating or 'stealth seed' networks. Yet, Herring also points out that the appropriation of the technology by farmers may ultimately accelerate the commercial deployment of the sterility-producing 'Terminator' gene, which became a symbol of international opposition to GM seeds in the 1990s. I would argue that the expansion and commercial potential of biotech crops in developing countries makes the broad enforcement of patent rights through legal and biological mechanisms an increasingly likely corporate strategy. It is thus essential for future research to examine the contestation and negotiation of agricultural biotechnology by farmers, social movements and governments within a potentially shifting terrain of patent enforcement.

A common set of themes identified in an appraisal by the United Nations, the World Bank and the World Health Organization is three-fold: an emphasis on the potential and need for agricultural biotechnology innovation to meet future food demands; a recognition of the technology's contemporary limitations for supporting small-scale farming systems as rooted in private versus public control of the technology; and the possible continuation if not exacerbation of economic inequality with the expansion of biotech crops in large-scale industrialized farming systems (IAASTD 2009). This account thus calls attention to the underlying political economic context in describing the potential limitations of the Gene Revolution to address contemporary and projected food insecurity.

Rather than viewing a problematic future for agricultural biotechnology as pre-determined, policy discussions are critically envisioning ways to refashion a pro-poor, pro-producer Gene Revolution through the expansion of public research and the creation of farmer-saved biotech seed varieties (The Royal Society 2009). For academic research, the task is to examine how institutional discussions are actually unfolding and who has a voice in the process. Continued attention to political struggles over biotechnology knowledge offers the potential to engender more progressive policy guided by the practices and needs of small-scale farmers in the promotion of more sustainable agriculture. The degree to which future biotechnology discussions incorporate a range of perspectives may ultimately determine if the next proposed agricultural revolution will prove beneficial or undermine domestic food security, biodiversity and rural livelihoods.

Acknowledgements

I would like to thank Elizabeth Oglesby, Marv Waterstone, Paul Robbins, Robin Lewis, Pratyusha Basu, Bruce Scholten, Michael Steinberg, Juanita Sundberg and Gerardo Otero for their insights on various versions of this article. Thank you to Matthew Creelman, Catalina Lopez, Norma Maldonado and representatives from REDSAG for their support in Guatemala. Finally, I would like to express my gratitude to Fulbright-Hays, the National Science Foundation and the Charlotte Newcombe Foundation for financial support provided in the realization of my doctoral research and dissertation.

References

Abramson, E., 2009. Soy: a hunger for land. *NACLA Report on the Americas*, 42 (3), 34–41.

Alteri, M., 2009. Agroecology, small farms, and food sovereignty. *Monthly review*. Available from: www.monthlyreview.org/090810altieri.php [Accessed 20 May 2010].

Aquino, P., Carrión, F., Calvo, R., and Flores, D., 2000. Selected maize statistics. *In*: P. Pingali, ed. *CIMMYT 1999–2000 world maize facts and trends. Meeting world maize needs: technological opportunities and priorities for the public sector*. Mexico: CIMMYT, 45–57.

Arias, A., 1990. Changing Indian identity: Guatemala's violent transition to modernity. *In*: C. Smith, ed. *Guatemalan Indians and the State: 1540–1988*. Austin, TX: University of Texas Press, 230–237.

Bailey, R., 2005. *Liberation biology: the scientific and moral case for the biotech revolution*. New York: Prometheus Books.

Barlett, D., Steele, J., 2008. Monsanto's Harvest of Fear, *Vanity fair*. Available from: www.vanityfair.com/politics/features/2008/05/monsanto200805 [Accessed 15 August 2009].

Bebbington, A., 2000. Reecountering development: livelihood transitions and place transformations in the Andes. *Annals of the association of american geographers*, 90 (3), 495–520.

Binimelis, R., Pengue, W., and Monterroso, L., 2009. "Transgenic treadmill": responses to the emergence and spread of glyphosate-resistant johnsongrass in Argentina. *Geoforum*, 40 (4), 623–633.

Boyd, W., 2003. Wonderful potencies: deep structure and the problem of monopoly in agricultural biotechnology. *In*: R. Schurman and D. Kelso, eds. *Engineering trouble: biotechnology and its discontents*. Berkeley: University of California Press, 24–62.

Bulmer-Thomas, V., 1987. *The political economy of Central America since 1920*. Cambridge: Cambridge University Press.

Carey, D., 2009. Guatemala's green revolution: synthetic fertilizer, public health and economic autonomy in the Mayan highland. *Agricultural history*, 83 (3), 283–322.

CEH, 1999. Guatemala Memoria del Silencio. *Comisión para el Esclaracimiento Histórico*. Available from: www.shr.aaas.org/guatemala/ceh/mds/spanish/toc.html [Accessed 1 March 2011].

Chandler, S. and Dunwell, J., 2008. Gene flow, risk assessment and the environmental release of transgenic plants. *Critical reviews in plant sciences*, 27 (1), 25–49.

Conway, G., 2003. *From the green revolution to the biotechnology revolution: food for poor people in the 21st century*, Woodrow Wilson International Center for Scholars, Directors Forum. Available from: www.unc.edu/courses/2006spring/geog/021/001/Conway.pdf [Accessed 6 April 2009].

Cristiani, A., 2007. *Biotecnologia en maices del trópico*. Guatemala: Feria Nacional de Maíz. Retalthuleu.

DeSantis, S., 2003. *Control through contamination: US forcing GMO maize and free trade on Mexico and Central America*. Institute for Social Ecology Biotechnology Project/ACERCA. Available from: www.tradeobservatory.org/library.cfm?refID=25990 [Accessed 6 June 2007].

Dunkerley, J., 1988. *Power in the Isthmus: a political history of modern Central America*. London: Verso.

The Economist, 2009. *Monsanto: the parable of the sower*. Available from: www.economist.com/displaystory.cfm?story_id=14904184 [Accessed 25 November 2009].

Essex, J., 2008. Biotechnology, sound science, and the foreign agricultural service: a case study in neoliberal rollout. *Environment and planning C: government and policy*, 26 (1), 191–209.

Fitting, E., 2011. *The struggle for maize: campesinos, workers, and transgenic corn in the Mexican countryside*. Durham and London: Duke University Press.

FAOSTAT, 2011. Food and Agriculture Organization. Available from: faostat.fao.org/site/339/default.aspx [Accessed 25 April 2011].

Foresight, 2011. *The future of food and farming (2011) final project report*. London: The Government Office for Science.

Fradejas, A., Alonzo, F., and Durr, J., 2008. *Caña de azúcar y palma africana: Combustibles para un nuevo ciclo de acumulación y dominio en Guatemala*. Guatemala City, Guatemala: Coordinación de ONG y Cooperativas (CONGCOOP).

Fuentes, M., Van Etten, J., Vivero, J., and Ortega, A., 2005. *Maiz para Guatemala: Propuesta para la Reactivacioón de la Cadena Agroalimentaria del Maíz Blanco y Amarillo*. Guatemala City, Guatemala: Food and Agriculture Organization (FAO).

Gauster, S., 2007. Eliminating market distortions, perpetuating rural inequality: an evaluation of market-assisted land reform in Guatemala. *Third world quarterly*, 28 (8), 1519–1536.

Giovannetti, M., 2003. Ignoring complex interactions in natural ecoysystems: the case of terminator technology. *In*: V. Benci, P. Cerrai, P. Freguglia, G. Israel and C. Pellegrinieds, eds. *Determinism, holism, and complexity*. New York: Kluwer Academic/Plenum Publishers.

Grandin, G., 2004. *The last colonial massacre: Latin America in the cold war*. Chicago: University of Chicago Press.

GRAIN, 2010. Landgrab resource page. Available from: www.grain.org/landgrab/ [Accessed 20 March 2011].

Hale, C., 2002. Does multiculturalism Menace? Governance, cultural rights and the politics of identity in Guatemala. *Journal of Latin American studies*, 34, 485–524.

Handy, J., 1984. *Gift of the Devil: a history of Guatemala*. Boston: South End Press.

Herring, R., 2007. Stealth seeds: bioproperty, biosafety, biopolitics. *Journal of development studies*, 43 (1), 130–157.

Huete, S., 2005. *Análisis sobre la 'Propuesta de Ley de Seguridad de Biotecnología Moderna para Guatemala*. Guatemala City, Guatemala: USDA Office of Agricultural Affairs – American Embassy.

IAASTD, 2009. *Agriculture at a crossroads: volume III Latin America and the Caribbean. International assessment of agricultural knowledge, science and technology for development*. Washington, DC: Island Press.

James, C., 2010. *Global status of commercialization biotech/GM crops: 2010*. Ithaca, New York: International Service for the Acquisition of Agri-Biotech Applications (ISAAA).

Jansen, K. and Roquas, E., 2008. Biosafety regulation and global governance: the problem of absentee expertise in Latin America. *In*: G. Otero, ed. *Food for the few: neoliberal globalism and biotechnology in Latin America*. Austin, TX: University of Texas Press, 91–113.

Jasanoff, S., 2005. *Designs on nature: science and democracy in Europe and the United States*. Princeton: Princeton University Press.

Jonas, S., 2000. *Of centaurs and doves: Guatemala's peace process*. Oxford: Westview Press.

Kardwell, K., 2005. *DRAFT-Embassy Science Fellowship-Guatemala: final report-assessment of the Guatemalan proposed framework for the regulation of biotechnology*. Guatemala: US State Department, USDA-FAS.

Klepek, J., 2007. *Falta de regulación facilitaría cultivo de transgénicos, Inforpress Centroamerica*. Available from: www.albedrio.org/htm/articulos/j/jk-001.htm [Accessed 20 June 2008].

Kloppenburg, J., 2005. *First the seed: the political economy of plant biotechnology, 1492–2000*. New York: Cambridge University Press.

Levidow, L. and Boschert, K., 2008. Coexistence or contradiction? GM crops versus alternative agricultures in Europe. *Geoforum*, 39, 174–190.

Manz, B., 2004. *Paradise in ashes: a guatemalan journey of courage, terror and hope*. Berkeley: University of California Press.

McAfee, K., 1999. Selling nature to save it? Biodiversity and green developmentalism. *Environment and planning: society and space*, 17 (2), 133–152.

McAfee, K., 2003. Neoliberalism on the molecular scale. Economic and genetic reductionism in biotechnology battles. *Geoforum*, 34, 203–219.

McAfee, K., 2008a. Exporting crop biotechnology: the myth of molecular miracles. *In*: G. Otero, ed. *Food for the few: neoliberal globalism and biotechnology in Latin America*. Austin, TX: University of Texas Press, 61–90.

McAfee, K., 2008b. Beyond techno-science: transgenic maize in the fight over Mexico's future. *Geoforum*, 39, 148–160.

McCarthy, J. and Prudham, S., 2004. Neoliberal nature and the nature of neoliberalism. *Geoforum*, 35, 275–283.

Mellon, M. and Rissler, J., 2004. *Gone to seed: transgenic contaminants in the traditional seed supply*. Cambridge, MA: Union of Concerned Scientists.

Mercer, K. and Wainwright, J., 2008. Gene flow from transgenic maize to landraces in Mexico: An analysis. *Agriculture, Ecosystems and environment*, 123, 109–115.

Mohr, K. and Tebbe, C., 2007. Field study results on the probability and risk of a horizontal gene transfer from transgenic herbicide-resistant oilseed rape pollen to gut bacteria of bees. *Applied genetics and molecular biotechnology*, 75, 573–582.

Monsanto, 2011. Fourth-Quarter 2011 Financial Results. Available from: www.monsanto.com/investors/Documents/2011/FINAL-Q4_Monsanto-Earnings_Slides-10-05-11.pdf [Accessed 5 November 2011]

Ochoa, L., 2009. Crisis alimentaria a flote, debilidad institucional al fondo. 1818. Infopress Centroamericana. Available from: www.inforpressca.com [Accessed 10 April 2010].

Oglesby, E., 2004. Corporate citizenship? Elites, labor and the geographies of work in Guatemala. *Environment and planning D: society and space*, 22, 553–572.

Otero, G., ed., 2008. Neoliberal globalism and the biotechnology revolution: economic and historical context. *In: Food for the few: neoliberal globalism and biotechnology in Latin America*. Austin, TX: University of Texas Press, 1–29.

Otero, G. and Pechlaner, G., 2009. Is biotechnology the answer? The evidence from NAFTA. *NACLA report on the Americas*, 42 (3), 27–31.

Piñero-Nelson, J., van Heerwaarden, J., Perales, H., Serratos-Hernández, J., Rangel, A., Hufford, M., Gepts, P., Garay-Arroyo, A., Rivera-Bustamante, R., and Álvarez-Buylla, E., 2009. Transgenes in Mexican maize: molecular evidence and methodological considerations for GMO detection in landrace populations. *Molecular ecology*, 18, 750–761.

PNUD, 2010. *Guatemala: hacia un estado para el desarollo humano: informe nacional de desarollo humano 2009/2010*, Programa de las Naciones Unidas para el Desarollo.

Quist, D. and Chapela, I., 2001. Transgenic DNA introgressed into traditional maize landraces in Oaxaca, Mexico. *Nature*, 414, 541–543.

Robbins, P., 2004. *Political ecology: a critical introduction*. Oxford: Blackwell Publishing.

The Royal Society, 2009. *Reaping the benefits: science and the sustainable intensification of global agriculture*. London: The Royal Society.
Schurman, R. and Munro, W., 2010. *Fighting for the future of food: activists versus agribusiness in the struggle over biotechnology*. Minneapolis: University of Minnesota Press.
Schwartz, N., 1990. *Forest society: a social history of Peten, Guatemala*. Philadelphia: University of Pennsylvania Press.
SENACYT, 2007. *Symposium on science-based biotechnology regulatory policy*. Guatemala City, Guatemala: National Secretariat of Science and Technology.
Shiva, V., 1991. *The violence of green revolution: third world agriculture, ecology and politics*. London: Zed Books.
Shiva, V., 2001. *Protect or plunder? Understanding intellectual property rights*. London: Zed Books.
Solano, L., 2008. Reconversión productive y agrocombustibles. *El Observador*, 3 (14), 31–61.
Soleri, D., Cleveland, D., Aragón, F., Fuentes, M., Ríos, H., and Sweeney, S., 2005. Understanding the potential impact of transgenic crops in traditional agriculture: maize farmers' perspectives in Cuba, Guatemala and Mexico. *Environmental biosafety research*, 4 (3), 141–166.
Soleri, D., Cleveland, D., and Cuevas, F., 2006. Transgenic crops and crop varietal diversity: The case of maize in Mexico. *Bioscience*, 56 (6), 503–513.
Streeter, S., 2006. Nation-building in the land of eternal counter-insurgency: Guatemala and the contradictions of the Alliance for Progress. *Third world quarterly*, 27 (1), 57–68.
Tay, K., 2009. *Guatemala: biofuels annual, biodiesel and ethanol*, USDA Foreign Agricultural Service GAIN Report, GT9008. Available from: www.gain.fas.usda.gov/.../General%20Report_Guatemala_Guatemala_5-26-2009.pdf [Accessed 10 August 2009].
Teubal, M., 2008. Genetically modified soybeans and the crisis of Argentina's agriculture model. *In*: G. Otero, ed. *Food for the few: neoliberal globalism and biotechnology in Latin America*. Austin, TX: University of Texas Press, 189–216.
Tobis, D., 1976. Las corporaciones norteamericanas más importantes de Guatemala. *In*: S. Jonas and D. Tobis, eds. *Guatemala: Una Historia Inmediata*. Mexico: Siglo XXI, 263–288.
Van Etten, J., 2006. Molding maize: the shaping of a crop diversity landscape in the western highlands of Guatemala. *Journal of historical geography*, 32, 689–711.
Van Etten, J. and Fuentes, M., 2004. 'La Crisis Del Maíz en Guatemala: Las Importaciones de Maíz y la Agricultura Familiar', *Anuario de Estudios Centroamericanos*, 30 (1–2), 51–66.
Van Etten, J. and de Bruin, S., 2007. Regional and local maize seed exchange and replacement in the western highlands of Guatemala. *Plant genetic resources: characterization and utilization*, 5, 57–70.
Vilas, C., 1995. *Between earthquakes and volcanoes; markets, state, and revolutions in Central America*. Monthly Review Press, New York.
Yapa, L., 1996. Improved seeds and constructed scarcity. *In*: R. Peet and M. Watts, eds. *Liberation ecologies: environment, development, social movements*. London: Routledge, 69–85.

Evaluating the Green Revolution after a decade: a Swaziland case study

Alan Terry

Department of Geography and Environmental Management, University of the West of England, Bristol, Frenchay Campus, Coldharbour Lane, Frenchay, Bristol BS16 1QY, UK

This paper analyses the impact of the Komati Pilot Project (KPP) upon members and their neighbours. The KPP was a Government of Swaziland (GoS)-financed irrigation scheme that enabled 43 subsistence and semi-subsistence farmers to change from rain-fed maize and cotton to irrigated sugar cane. It provides an opportunity to evaluate the impact of Green Revolution technology upon an African rural community where adopters and non-adopters live close together and where little income differentiation existed prior to the development because the common experience was poverty. A baseline survey of the KPP was carried out in 1997 in advance of the Komati Downstream Development Project. Of concern are the extent and direction of any trickle-down effects from the KPP participants for their neighbours and the consequences of any such impacts 10 years after the project started producing sugar cane. The expectation of the GoS was that sugar cane would benefit KPP participants and neighbours creating a multiplier effect through increased demand for services and increased demand for agricultural labour. Ten years later, these expectations are unrealized.

The Green Revolution in an African context

Throughout the first decade of the 21st century, there has been a rediscovery of the central importance of agriculture in the development process, especially in addressing the key issue of poverty reduction (FAO/World Bank 2001, IFAD 2001, Irz *et al.* 2001, Dorward *et al.* 2004). As part of this process, the World Bank's (2008, p. 1) call for a revolution in the productivity of African agriculture acknowledges that this has to be different from the Asian Green Revolution 'due to Sub-Saharan Africa's unique agriculture and institutions'. When asking the question 'Why Has Africa Not Experienced an Asian-type Green Revolution?' Agwe *et al.* (2007) argue that it is due to more difficult agro-climatic conditions such as variable rainfall and poorer soil fertility and poorer access to input or output markets. The World Bank (2006a, 2006b) cites low and stagnant labour productivity, poor health and educational levels of farmers and lack of mechanized farming. A further factor is the lack of a supportive environment in key institutions. This sidesteps two issues. First, has the Green Revolution been introduced at a smaller scale within Africa and second, has its introduction been desirable? With respect to the first question, Biello (2007) argues that rising African agricultural productivity between 2000 and 2005 was proof that a revolution was already underway with targeted aid and better technologies beginning to reduce rural poverty as a result of grass-roots-type initiatives such as the Millennium Villages Project. With respect to the second question, Niazi (2004) concludes that in Pakistan the Green Revolution has widened income differentials between the rich and the poor, because the technology is already biased towards the rich and against the poor, creating a virtuous cycle for the former and a vicious cycle for the latter. In contrast, Estudillo (2006) provides evidence that in the Philippines, even the poorest landless individuals benefitted from the introduction of the Green Revolution because the demand for rural labour increased the real wages of unskilled rural farm workers. Shiva (2008) emphasizes the unsustainability of Green Revolution technology due to dwindling oil supplies and rising energy prices. In Africa, such arguments have become salient given the optimism expressed by the promoters of initiatives such as the Alliance for a Green Revolution in Africa, as the technology is seen as a solution to improving food security and reaching Millennium Development Goals (MDGs) of poverty alleviation. There appears to be little questioning of the benefits of this course of action or appreciation of potential problems. However, most of the experience of Green Revolution technology and its impact has been outside Africa and relatively little empirical evidence exists within African communities. Xinshe and Headey (2008) provide theoretical evidence based on economic modelling that, if Green Revolution technology were to be applied to staple production, consumers and producers would benefit and 70 million Africans would be lifted out of poverty, although no mention is made of the equity effects. A key element for the successful introduction of Green Revolution technology is irrigation (Economic Commission for Africa 2003, Polak 2005). In this respect, the Southern African Development Community's (SADC) Regional Indicative Strategic Development Plan (2003) aims at doubling the area under irrigation by 2015 from a very low 1.4 per cent of arable land (African Development Bank 2011), recognizing the importance of improving agricultural productivity in a region which is dominated by the agricultural sector and where raising rural incomes is a key development goal in tackling chronic levels of poverty (Chilonda *et al.* 2007).

Swaziland, one of SADCs 14 members, has the highest percentage of irrigated agricultural land in southern Africa, although most is on title deed land (TDL) (Funnell 1991). Since the late 1990s, it has attempted to improve agricultural productivity on traditionally tenured Swazi Nation Land (SNL) by extending irrigated sugar cane to these areas. Like much of sub-Saharan Africa, the impact of Green Revolution technologies on Swaziland's traditional agricultural sector has been minimal compared to those in south and south-east Asia and Latin America. Although hybrid maize, the main staple, occupies more than 80 per cent of cropped area on SNL (Van Waveren 2006), its impact on food security and national food production has been negligible as less than 5 per cent is irrigated and fertilizer use is low (Terry 2001, Van Waveren 2003). Rainfall is unreliable, especially in the semi-arid low veld and lower-middle veld, with climatic drought (two or more consecutive years with rainfall well below average) being a regular feature (Gommes 1996 cited in Van Waveren 2006, p. 206). Recently, the frequency and severity of droughts appear to be increasing (Rouault and Richards 2003). The vulnerability to drought is increased because maize is not as drought resistant as millet or sorghum and it has been estimated that more than 50 per cent of the country falls outside the potentially successful rain-fed growing areas for the crop (Van Waveren *et al.* 1994). The effect of variable weather on maize production is illustrated by the drought of 1991/92, where productivity on SNL fell by 62 per cent from the previous growing season (World Bank 2000). This helps to explain why even in good rainfall years, despite the fact that 70 per cent of the population is engaged in agriculture, 60 per cent of the cereal requirement is imported (United Nations Development Programme 2006).

Within Swaziland, Green Revolution technology has been confined mainly to TDL, 25 per cent of the country, via the introduction of irrigated sugar cane, but since the late 1990s has become the main vehicle for modernizing traditional Swazi farm systems. Since then, irrigation has been extended to SNL, enabling Green Revolution technology to spread to traditional farming communities. However, these are already differentiated with respect to income (Funnell 1991) and so if sceptics are correct (Cleaver 1972, Gadgil and Guha 1995, Niazi 2004, Shiva 2008) there may be a widening of inequalities unless Swaziland's socio-economic circumstances provide opportunities for Green Revolution technologies to spread wealth and reduce Swaziland's chronic rural poverty.

The Kingdom of Swaziland's agricultural policy

The landlocked Kingdom of Swaziland, the second smallest country in Africa, has a population of 1.1 million, 76 per cent rural. In 2003, its per capita income of US$1,350 was lower middle, but the poorest 60 per cent had an average lower than that of sub-Saharan Africa (Government of Swaziland 2001) and the Gini coefficient of 0.60 gave it one of the most skewed income distributions in the world (UNDP 2006). Sixty-nine per cent lived below the income poverty level and 37 per cent in extreme poverty (i.e. below the food poverty line). In rural areas, the figures were 76 and 43 per cent (Government of Swaziland 2001), but poverty is concentrated on SNL where the majority of the population rely for all or part of their incomes on low-productivity agriculture (Government of Swaziland 1999, Atkins 2007). Until 1999, a principal goal of agricultural policy had been to achieve food security through self-sufficiency in maize production, mostly from SNL. Increased agricultural exports were seen as the role of large-scale privately owned mainly irrigated estates producing sugar cane and fruit on TDL. The policy resulted in a wide productivity gap between SNL and TDL and chronic food insecurity for a significant proportion of the population (World Bank 2003).

In 1999, Government of Swaziland (GoS) moved towards a new agricultural policy that attempted to increase food security through the commercialization of the smallholder sector. The key to raising SNL productivity is irrigation, to offset unreliable rainfall. In order to achieve its goals, in the late 1990s, GoS formulated two irrigation projects: the Komati Downstream Development Project (KDDP) enabling 1,500 homesteads on 6,000 hectares of SNL to be irrigated in the northern semi-arid low veld of Swaziland, and the Lower Usuthu Smallholder Irrigation Project (LUSIP), which when completed will irrigate 1,750 homesteads on 11,500 hectares in the central low veld.

The case study: the Komati Pilot Project

This paper aims to analyse the long-term economic impact of the Komati Pilot Project (KPP) sponsored by the GoS in the late 1990s to provide the Ministry of Agriculture and Cooperatives (MoAC) with information on the impact of small holder irrigated sugar cane on a rain-fed maize, predominantly subsistence community, prior to the development of the KDDP. The specific research questions are:

(1) To what extent had income changed for KPP members and non-members between 1997 and 2007?
(2) To what extent had wealth changed for both groups? Key wealth indicators include:
  - Livestock, specifically cattle because of their wider cultural significance,
  - Consumer goods and vehicles,
  - Housing.
(3) To what extent had food security changed?
(4) How did both groups perceive the impact of the KPP on:
  - Their families?
  - The community?

The KPP covered two areas, one in the Komati valley and the other in the Usuthu valley. Farmers in the Komati are members of the Nyakatfo Farmers' Association, and those in the south, The Lobovu Farmers' Association. Because of resource constraints, the 2007 follow-up survey only visited Nya-katfo. In association with MoAC, the author carried out a baseline survey of the KPP in 1997 and the results of this together with lessons learned from the initial stages of the project indicated a number of issues that were taken into consideration on the KDDP (Terry 1997, 2001). The KPP therefore provides an opportunity to evaluate the impact of a Green Revolution technology (a bundle of agricultural inputs that include improved seeds, intensive use of chemical fertilizers, pesticides, herbicides and irrigation) upon an African rural community. In this case, adopters and non-adopters live close together and little income differentiation existed prior to the development. The common experience before the KPP was poverty.

The implications of the research are of national importance because individual farmers' associations are still being formed on the KDDP and LUSIP, and the extent to which resources are shared equitably varies greatly. If assumptions regarding trickle-down are wrong, then given the goal of poverty alleviation that lies at the heart of both KDDP and LUSIP, this needs to be addressed with some urgency. However, given that SADCs aim to achieve double-irrigated agriculture by 2015, to reduce poverty reduction through improving agricultural productivity, the findings are also of regional importance because claims are being made for similar schemes such as SADCs Upper Okovango Basin Project in Namibia and Angola which is projected to reduce poverty and improve food security for the participating countries, as well as 200,000 people in the neighbouring countries of Zambia, Zimbabwe and Botswana (African Development Bank 2011). Similar arguments are also being put forward to support what Hall (2011) terms as the 'Next Great Trek' by South African and foreign-backed commercial agribusinesses to the wider SADC region and beyond. In the light of the debate concerning the impact of a trickle-down effect on such development projects, the empirical evidence from the KPP follow-up survey provides evidence for the extent, direction and nature of any such effects on both project participants and their immediate neighbours who have never been directly involved in the project and therefore any such longitudinal studies at a local scale would appear to be salient at this time.

Trickle-down effects

Of particular concern in this paper is the analysis of the extent and direction of any trickle-down effects from the project members to their non-participating neighbours and the consequences of any such impacts 10 years after the KPP started producing sugar cane. The expectation by MoAC was that benefits would accrue to both project participants and non-participants (Terry 2001). The main crop would be sugar cane and the increased income for farmers would enable a multiplier effect in income and employment into surrounding homesteads, as it was expected that the demand for services such as transport, agricultural machinery and farm labour would benefit those with little direct involvement in the irrigation projects.

A major consideration for development practitioners and theorists is the likelihood of a trickle-down effect permeating from development poles into surrounding areas. Hirschmann's (1958) optimistic view was that uneven growth would eventually percolate into all areas of the economy from the global to national scale and that imbalances in the development process were inevitable and to a certain extent desirable. This model, similar to that of Freidmann's (1966) core–periphery model, became the basis for many early development strategies whose basic assumption was that all societies were on essentially the same development path towards modernization (Potter *et al.* 2004). However, such approaches have been criticized for oversimplifying underlying historical, social and cultural influences, while 'the trickle-down process has often failed to materialise among those who have been the subject of modernisation projects' (Daniels *et al.* 2008, p. 188). Chenery *et al.* (1974) were among the first to highlight the failure of these approaches and argued that development strategies would need to consider more attempts targeted at poverty alleviation, arguing for policies based on growth with redistribution as a key goal. In the wake of the twin oil shocks of 1974 and 1979 and the rising Debt Crisis, the neoliberal Washington Consensus marginalized considerations of poverty alleviation and equity. Although the effects of structural adjustment policies are still contested, by the mid-1990s it was becoming increasingly obvious that they had created and enhanced inequalities at all scales, with some arguing that this may well have been a policy goal (George 2007). Although adoption of the MDGs and the G8 Gleneagles Agreement (2005) have raised the profile of poverty reduction, it is highly unlikely that targets will be met, especially in Africa.

At the more focused rural development level, Little and Watts (1994) and Bernstein (1992) suggest that dependency and differentiation of rural communities are a tendency of commercialization of traditional farming, leading to negative effects on food security for the poor. Key and Runsten (1999) suggest that contract farming has benefited large-scale compared to small-scale farms, and Porter and Phillips-Howard (1997) argue that small-scale farmers might benefit, although those they employ would not. Stringfellow (1996) concluded that problems of food security are likely to be worsened where land is in short supply, or where planners fail to acknowledge existing risk management strategies of the farmers. Food First critics argue that cash crops undermine food security by cutting down subsistence production, which market relations fail to replace (George 1976, Lappe and Collins 1988). In practice, the distinction is less obvious. In a survey of the research literature on cash crop development schemes and nutrition, Longhurst (1988) reported positive and negative effects. However, Govereh and Jayne (2003) argue that cash crops have synergistic impacts on subsistence food production. Research in the Gokwe cotton-producing region of Zimbabwe revealed that cotton producers gained access to credit and management training that in turn led to higher yields of subsistence crops than their neighbours who did not produce cotton, although the multiplier effects of cotton production were also felt among non-cash crop producers. The initial impact of the KPP was very positive on the participants, and by 2001 the Nyakatfo Farmers' Association (KPP members) had repaid all loans. This was within three years of the start of farming operations (Komati Downstream Development Project 2001), although, the average subsidy of US$17,895 per KPP homestead was much too high to be extended to all the KDDP associations then beginning to form.

Methodology

Between April and July 1997 a baseline survey of 154 homesteads, of which 43 were KPP members, was undertaken by a team of five MoAC enumerators. This represented 80 per cent of the KPP members and 90 per cent of all the homesteads in the Nyakatfo area. The methodology was a formal questionnaire in which information relating to the following issues was collected:

(1) demographic characteristics;
(2) residential status;
(3) reliance on wage income;
(4) educational attainment of
  - all homestead members,
  - homestead heads and or members of the associations (decision makers);
(5) farm size;
(6) cattle ownership;
(7) wealth indicators including cattle.

In June and July 2007, 67 homesteads from the original 1997 survey were sampled, 19 members of the KPP, comprising 28 per cent of the survey, the same as the baseline survey. The discrepancy in total number surveyed was due to the difference in resources available for the two surveys. However, as the proportion of members to non-members was the same in both surveys, this provided a comparison. Respondents were sampled to provide the same farm size distribution as in 1997 (see Table 2). The 1997 questions were repeated so that a direct comparison could be made between the two sets of data. For both surveys, the head of homestead was interviewed, but where he or she was not available, another adult family member was substituted. Interviews were carried out in Siswati with translators employed from the University of Swaziland. All homesteads in the 2007 survey had been interviewed in the 1997 survey, and so any newly formed homesteads, of which there were very few, were not considered for inclusion in the follow-up.

A brief description of the Nyakatfo community

Nyakatfo is a typical dispersed Swazi rural settlement with no obvious central point. In 1997, it consisted of approximately 180 homesteads spread fairly evenly over an area of 10 km^2 on a river terrace to the north of the Komati River. There appeared to have been very little change in the number of homesteads between 1997 and 2007 with fewer than 10 being identified by the chief. The total sample in 1997 was 1,025 which was approximately 90 per cent of the total population. Twenty-seven per cent of the sample lived in KPP member homesteads. Forty-two per cent was under 16 years of age, with only 12 per cent over 45. There was no statistically significant difference in age, occupation, place of residence of heads of household or educational status between KPP members and their neighbours (Terry 2001). KPP members averaged eight persons per homestead while non-KPP homesteads averaged 6.4. In 2007, the total population in the sampled homesteads was 468 with an average of 8.2 per homestead for KPP members and 6.5 for non-members. No significant differences existed in population structure between 1997 and 2007, although a more detailed analysis and comparison of individual families would certainly have picked up changes reflecting family life cycles and probably the impact of the HIV/AIDS epidemic on specific age groups and individuals between the two periods. This was not possible because to ensure confidentiality, the original data sets did not relate individual homestead data to specific locations. KPP members are scattered randomly throughout the settlement and no spatial division exists between them and their neighbours. A common factor concerning KPP members was that they had all been members of a cotton co-operative that had ceased functioning a few years before the start of the KPP. Some non-KPP members had also been a part of this group, but had chosen not to enter the KPP when negotiations began between its committee and the government to switch from cotton to sugar cane.

Results and analysis

Income

In 1997, average per capita income of KPP members was E (Emelangani) 796 (US$278), while the homestead mean of the non-members was E743 (US$260). The majority of both groups were living in absolute poverty. By 2007, average incomes in KPP homesteads had risen to US$13,272 (US$1,659 per capita). In contrast, the majority of the non-members had witnessed declining or stagnant incomes over this period caused by loss of employment due to retrenchment in the sugar estates, or loss of farm-derived income caused by two droughts in 2002–03 and 2006–07. In 2007, over half of this group were dependent on a World Food Programme drought relief scheme. Because most homesteads had little or no formal income at the time of the second survey and the majority were dependent upon food aid, no accurate estimate could be made of their income.

In contrast to the process of de-agrarianization that has been a feature of much rural African life over the past 20 years (Bryceson 2002), the KPP has enabled its members to become re-agrarianized. On a livelihood analysis (Carney 1998, Scoones 1998), the key difference between the groups was the access to social capital via the cotton cooperative (Terry 2001). This created the forum for initial ideas to convert to irrigated sugar cane. Additional factors were a relatively young age group and the fact that 25 per cent of the group had recently lost their jobs. This nucleus had the drive, vision and initiative to galvanize and keep most of the group together over the long gestation period of the KPP. Many also had key literacy and numeracy skills and some had worked in the sugar industry. The group also contained individuals with the ability to lead and organize, which had raised aspirations as a consequence of having worked in South Africa. Friendship and family ties had further reinforced the trust within the group which was another vital factor before any commitment to change crops after the collapse of the cotton price in the early 1990s.

Table 1 shows that Nyakatfo Farmers' Association members already had higher incomes than their neighbours, but both groups were living below or around absolute poverty levels. However, these estimated incomes should be treated with caution because, if derived from the sale of surplus agricultural produce or the informal labour market, they are extremely difficult to quantify accurately because of a lack of record keeping and the highly variable nature of most transactions.

Farm size

Farm size categories are the same as those used in 1997 and earlier MoAC surveys. Non-members averaged 3.2 hectares while members averaged 4.9 hectares and no changes in farm size on sampled farms had occurred over this period (Table 2). In 1997, average farm size on SNL was 1.94 hectares (Terry 1997) but 3.61 hectares in the semi-arid low veld (Atkins 2007). Located in the northern lowveld, Nyakatfo's dependable rainfall of approximately 550 mm per annum, mostly falling between December and April, is at the margin for rain-fed maize.

In 1997, farmers tended to follow a relatively simple cropping pattern. Only 26 per cent grew more than five crops while 43 per cent grew a maximum of two crops; 40 per cent of this latter group grew only maize which dominated both sub-groups, accounting for 60 per cent of the cropped area. Rain-fed cotton was the second most important crop. All the farmers' association members had grown it and it was this land, previously under cotton, that had been allocated to irrigated sugar cane. In contrast, only 22 per cent of non-members grew cotton. Differences in cotton-derived income were the major reason for differences in income from agriculture between the two groups. KPP members averaged E1,278 per annum from raw cotton sales compared to E275 for non-members. There had been no significant change in farm size or the nature of

Table 1. Estimated average homestead income from all economic activities: 1997 (Emelangeni (E) per annum).

	Nyakatfo Farmers' Association	Nyakatfo non-members
On farm	2,219	1,223
Off-farm	4,150	3,535
Total	6,369	4,758

Note: Average homestead income southern low veld: E5,640 (Atkins and Dlamini, 1998) US$1 = 2.86 Emelangeni: 1997 rate.

Table 2. Farm size (per cent of sample).

	Nyakatfo Farmers' Association		Nyakatfo non-members	
Hectares	1997	2007	1997	2007
0–0.99	0	0	20.9	20.8
1–1.99	0	0	27.8	27.1
2–2.99	35.1	36.8	26.4	25.0
3–5.99	48.7	47.4	8.9	10.4
≥6.0	16.2	15.8	16.1	16.7
Total (%)	100	100	100	100

Note: All respondents claimed that there had been no change in the size of their farms between 1997 and 2007.

the crops grown on individual farms over the 10-year period between surveys, except that KPP members each produced 2.2 hectares of irrigated sugar cane which now contributed the bulk of their income, while due to the drought, virtually no rain-fed crops had been produced by either sub-group in the previous growing season.

Cattle

Cattle are a traditional store of wealth and are important culturally as they are used for lobovo (bride price). In 1997, KPP members averaged 11.3 cattle per homestead compared with 10.3 for non-members. However, 34 per cent of both groups owned no cattle. By 2007, cattle numbers had fallen from 2,385 to 325. Given the smaller sample size, the expected herd would have been 1,073. The fall can be explained partly by the impact of two droughts, but the biggest percentage fall is in the KPP group with only 19 per cent of the expected herd remaining while for non-members the figure is 73 per cent. This indicates how KPP members are holding a greater percentage of their wealth in non-cattle assets, a significant cultural change between 1997 and 2007. The number of other livestock, goats and poultry was insignificant for both groups. As well as the arable land that farmers fence and utilize as private property, they also have access to communal range land used for grazing and collection of other resources such as firewood, clay and wild foods.

Wealth

A wealth index was created based on 48 indicators approved by The GoS. These ranged from household goods and vehicles to farm implements and animals. The change in wealth distribution within the sub-groups can be seen in Table 3. In 1997, KPP members were almost twice as likely to be in the wealthiest category as non-members and almost four times less likely to be in the least wealthy category. However, many relatively wealthy homesteads were not members of the KPP.

Table 3. Wealth distribution (per cent).

Wealth category	Nyakatfo Farmers' Association 1997	Nyakatfo Farmers' Association 2007	Nyakatfo non-members 1997	Nyakatfo non-members 2007
Wealthiest	50	53	30	17
Mid-wealth	39	47	27	25
Least wealthy	11	0	43	58
Total (%)	100	100	100	100

Ninety-two per cent of the bicycles were on non-KPP homesteads, and they owned the only cars in the community. Although the KPP seems wealthier, even very poor households were part of the group with 11 per cent of members appearing there.

By 2007, none of the KPP members remain in the least wealthy category, while there has been a fall of 13 per cent among non-members in the wealthiest category. This is mainly accounted for by the drop in cattle. In addition, there had been a substantial increase in the percentage of this group in the least wealthy category. Significant differences can now be seen in the ownership of consumer goods, access to electricity and vehicles.

Table 4 reveals that increases in ownership of all wealth categories are striking for the KPP, but modest for the non-participants. Noticeable differences occur in the most expensive assets such as light pick-ups and cars. Smaller items have a smaller differential, while the large percentage increase in telephones for both groups reflects the introduction and decline in the price of cell phones over the period.

Although Table 4 demonstrates the growing discrepancy in wealth, some spread has occurred. However, the speed and difference in the growth of wealth between the two groups are unprecedented, especially given that the majority had lived at or near absolute poverty in 1997. These changes do not show the full change in wealth as they exclude housing. Eighty-four per cent of KPP members had constructed a new main dwelling in the past 10 years compared to 56 per cent of non-members. Traditionally, Swazis live in one-room rondavels with different groups having access to different buildings in a single homestead. Many of the earliest modern housing simply replace traditional rondavels with concrete block, tin-roofed single-roomed structures. However, as incomes have grown, multi-roomed bungalows are constructed. This has been striking among KPP members since 1997.

Seventy-seven per cent of new homes on non-KPP members' homesteads are made up of relatively simple one- or two-roomed buildings (Table 5). In contrast, 63 per cent of new homes on KPP homesteads are three roomed or above. Many members have constructed modern brick-built bungalows with tiled roofs that would not look out of place in any suburban European setting. The improvement in housing quality, especially among KPP members, is one of the biggest changes and represents a major improvement in the quality of lives for those who have benefitted.

Food security

Given the contested nature of the impact of Green Revolution technology on food security, from proponents (Black 1960, Borlaug 1972) to critics (Cleaver 1972, Bernstein 1992, Little and Watts 1994, Gadgil and Guha 1995, Shrestha 2008), this factor will now be discussed. In 1997, KPP

Table 4. Key wealth indicators by sub-group (per cent).

Key indicator	Nyakatfo Farmers' Association 1997	Nyakatfo Farmers' Association 2007	Nyakatfo non-members 1997	Nyakatfo non-members 2007
Light pick-up	2	68	1	2
Lorry	2	0	0	0
Electricity	2	47	1	13
No formal education	53	53	71	69
Bicycle	2	47	22	21
Saloon car	0	21	2	2
Refrigerator	0	63	5	10
Television	2	63	3	21
Telephone	0	99	0	44

Table 5. Number of rooms per new dwellings (per cent).

Key indicator	Nyakatfo Farmers' Association	Nyakatfo non-members
1	19	46
2	19	31
3	19	15
4	31	0
5	13	8
Total (%)	100	100

members were only marginally better off with 41 per cent never buying maize compared to 37 per cent of their neighbours while 38 per cent purchased maize every year compared to 29 per cent of their neighbours. By 2007, the situation had reversed dramatically with 53 per cent of the KPP members attaining self-sufficiency every year while only 13 per cent of the non-KPP group were able to achieve this (Table 6).

A possible explanation, not explored in the survey, is that food security has been compromised more for non-KPP members than their KPP neighbours due to the differential impact of HIV/AIDS. The United Nations Development Programme (2006) states that the main cause of food insecurity in Swaziland is the impact of the HIV/AIDS epidemic. By 2005, it was estimated to have infected 38–40 per cent of the 15–49 age group making Swaziland the worst affected country in the world. The average life expectancy collapsed from 65 years in 1991 to 31 years in 2005 (Whiteside and Whalley 2007), 15 years below that attained in 1970 (UNICEF 2006) and much lower in comparison with the sub-Saharan Africa figure of 45.6 years. This has undermined the labour force necessary to produce subsistence crops, but where homesteads have access to adequate income, as is the case with the KPP members, they may be able to purchase labour to replace the drop in home-based labour caused by HIV/AIDS sickness or death. This has not been corroborated on this survey, but remains a potential explanation for the change in food security.

Female-headed households are particularly vulnerable (IFAD 2001). They comprise 22 per cent of the total sample, but 67 per cent of female-headed households are in the lowest wealth category compared to 34 per cent of male-headed households. None of this group, whether male or female headed, are members of the KPP. Only 7 per cent of female-headed homesteads are self-sufficient in maize every year compared to 28 per cent of male-headed households, and so this sub-group is more likely to need to purchase food but are less likely to have the means to do so than their male counterparts.

Summary of main changes to income and wealth in 2007

The conversion of 2.2 hectares per KPP member from rain-fed cotton to irrigated sugar cane raised net incomes from E252 (US$88) to E8,125 (US$2,841) per hectare (Terry 2001). Thus,

Table 6. Does the homestead buy maize? (per cent).

	Nyakatfo Farmers' Association 1997	Nyakatfo Farmers' Association 2007	Nyakatfo non-members 1997	Nyakatfo non-members 2007
Every year	38	16	29	52
Never	41	53	37	13
Occasionally	21	31	34	35
Total (%)	100	100	100	100

total income per KPP member from each 2.2 hectares was theoretically E16,375 (US$5,725). However, the KPP members had to repay loans received from MoAC (E1,663,350; US$587,591) and the Swaziland Development and Savings Bank (E2.2 million; US$769,230). These loans were repaid by the end of 2001 instead of the deadline of 2004 which indicates the profitability of the scheme and the transformative effect on KPP members, although in order to repay this amount dividends to individual members did not exceed E3,000 (US$1,049) per annum in the first three years. However, even this represented a 1.35-fold increase in on-farm income. Between 2001 and 2004, returns would have been nearer E16,375 (US$5,725) per member, a 7.7-fold increase in on-farm and a 2.6-fold increase in total income compared to the pre-KPP situation. This demonstrates the positive impact of the EU Sugar Protocol, which until recently raised the incomes of African Caribbean Pacific (ACP) sugar producers by an average of two to three times higher than free market prices.

In contrast, the majority of non-members witnessed declining or stagnant incomes caused by loss of employment, especially in the sugar estates due to retrenchment from 2002, or loss of income from the sale of surplus agricultural products caused by two devastating droughts in 2002–03 and 2006–07. Since 2001, the KPP members have benefitted from the Maguga Dam, which provides further insurance for their irrigation system in the face of possible longer and more frequent droughts.

How has the KPP been perceived by community members?

The impact of the KPP on the family

Given that one of the main intentions of the KPP was to stimulate multiplier effects, the fact that 10 years after the start of the KPP 80 per cent claimed that it had either no or a negative impact on them and their family suggests that this development model is challenged (Table 7). Only a minority of non-KPP homesteads have benefited and even this is limited, as most of the work they have managed to obtain has been part-time and therefore very poorly paid and seasonal. Only one homestead has opened a small business to service the expanding sugar cane industry by providing a vehicle repair service.

The majority of those who state that they are worse off have lost some land when it was converted from cotton to sugar cane. They had been members of the cotton cooperative, but did not join the KPP. Subsequently, they have been unable to recover their fields, or obtain compensation, either in the form of alternative land or cash. KPP members argue that anybody in the community may join the KPP, but the price to become a member is E40,000 (US$13,860), beyond the means of every non-KPP community member. This state of affairs is condoned by the local chief who remains the only source of power on SNL that could enable the former cotton cooperative members to regain their land. This demonstrates clearly the tenuous nature of entitlement to land on SNL and the need for individuals to strive to maintain use and access over it, even if

Table 7. How has the KPP impacted upon you and your family? (per cent).

	Nyakatfo Farmers' Association	Nyakatfo non-members
Positive	99	16
Mixed	1	4
No impact	0	60
Negative	0	20
No comment	0	0
Total (%)	100	100

Table 8. How has the KPP impacted upon the community? (per cent).

	Nyakatfo Farmers' Association	Nyakatfo non-members
Positive	68	25
Mixed	22	10
No impact	5	50
Negative	5	10
No comment	0	5
Total (%)	100	100

they are not fully in agreement with their neighbours about a change in its use. It also shows the level of ignorance that existed within many in the community about the transformative power of irrigated sugar cane.

The impact of the KPP on the community

KPP members are aware that their project has had a negative impact on their neighbours, both in terms of physically reducing the resources of some and through the huge and rapid discrepancy in personal wealth that has arisen. Large bungalows sit next to poorly maintained mud rondavels, many of which look close to collapse. Non-KPP members spoke of the rise in personal greed or the huge disparity in access to school through differences in the ability to pay school fees. They were, however, generally unwilling to criticize their KPP neighbours directly, possibly because they could see little that can be done to change the status quo because the KPP is supported by the chief. In addition, some former members of the cotton cooperative argue that they were cheated out of their share of the KPP because they were working away at the time when contracts were being signed. There is no doubt that the KPP has led to huge changes in the life chances of people who until recently had much more in common, summed up as shared poverty, although the rural Swazi society has always been hierarchical. However, the level of material disparity which has developed between 1997 and 2007 is new and, although some outside the KPP acknowledge the benefits, such as new roads and bridges, this traditional Swazi community is witnessing stresses caused by a rapid widening of wealth and incomes (Table 8).

Conclusions

Niazi (2004) argues that in Pakistan the Green Revolution has increased inequity, thereby harming development, the main reason being that the already rich were able to purchase the whole input package thereby increasing the value of land and pricing out tenants, especially on small farms. In Africa, this is unlikely to be as widespread because of customary tenured land. In Swaziland, access to SNL is a right and land cannot be sold because, even though farmers can fence and farm land they are allocated, they never own it. Other factors may create opportunities differentially within a community which Green Revolution technology then exacerbates. In the case of the KPP there were differences in social capital (Terry 2001). Membership of a failed cotton cooperative provided the original forum where individuals with the drive and vision to take a strategic view of their circumstances in the light of falling on- and off-farm incomes managed to persuade most of their cooperative partners to join the KPP. The government then failed to monitor the scheme carefully so that very large public subsidies enabled the conversion of virtually worthless cotton-cooperative membership into lucrative sugar cane membership. They have gained access to public funds to support a large-scale injection of capital and, due to the profitability of sugar while Swaziland was in receipt of premium ACP prices, were able to repay all loans within three years. However, as in Pakistan, increases in productive efficiency seem to have undermined social cohesion and the huge division between the haves and have-nots in a traditional rural society is of concern. This level of misallocation of funds has not occurred on the much larger KDDP which supports 10,500 people and evidence exists that some of the more recently formed farmers' associations have adopted more equitable systems (Terry and Ryder 2005, Terry 2007). However, given the very rapid formation of similar associations on LUSIP (12,000 people), the government must ensure equitable distribution of land and water rights within those communities that have not witnessed at close range the problems associated with an unequal share of the benefits of such a scheme.

Given the extra funding which is planned for African farming, these findings suggest that a priority should be an improvement in the capacity of those charged with distributing funds to beneficiaries to ensure equity, and also to ensure that individuals do not lose access to key resources during the transformation, a key concern (Hall 2011). While community participation approaches may enable this to occur, facilitators need to be aware of the ability of the well placed to capture and manipulate such processes for their own benefit (Cooke and Kothari 2002). This would help to reduce misallocation of support to relatively well-off community members, thereby preventing an exacerbation of pre-existing income and wealth inequalities as witnessed on the KPP. This remains a fundamental issue on customary tenured land throughout sub-Saharan Africa where traditional authorities retain great power and influence and reveals socio-political factors that are just as able to undermine sustainability as the more commonly recognized physical and economic factors.

References

African Development Bank, 2011. *SADC agriculture water management & food security – Upper Okavango Basin*. Available from: www.afdb.org/en/projects-and-operations/project-portfolio/project/p-z1-aac-006/ [Accessed 17 May 2011].

Agwe, J., Morris, M., and Fernandez, E., 2007. *Africa's growing soil fertility crisis: what role for fertilizer?* Agricultural and Rural Development Notes, 40398, Issue 21. Washington, DC: World Bank Available from: www.wds.worldbank.org/external/default/WDSContentServer/WDSP/IB/2007/07/30/000020953_20070730153304/Rendered/INDEX/403980 AFR0ARD01zer0note02101PUBLIC1.txt [Accessed 14 January 2009].

Atkins, S., 2007. *Investment decisions and rural poverty: an appraisal of the lower Usuthu Smallholder irrigation project in Swaziland*. Unpublished PhD thesis. Faculty of the Built Environment, University of the West of England, Bristol.

Atkins, S. and Dlamini, S., 1998. Results of a socio-economic baseline study of the proposed lower Usuthu Smallholder irrigation development project area: implications for project design. *Agrekon*, 38, 284–298.

Bernstein, H., 1992. Poverty and the poor. *In*: H. Bernstein, B. Crow and H. Johnson, eds. *Rural livelihoods: crises and responses*. Oxford University Press, 13–26.

Biello, D., 2007. Is a Green Revolution finally blooming in Africa? *Scientific American*, Available from: www.scientificamerican.com/article.cfm?id=is-green-revolution-finally-blooming-in-africa [Accessed 17 May 2011].

Black, E., 1960. *The diplomacy of economic development*. Cambridge, MA: Harvard University Press.

Borlaug, N., ed., 1972. *The Green Revolution*. Mexico: CIMMYT.

Bryceson, D., 2002. The scramble in Africa: reorienting rural livelihoods. *World development*, 30 (5), 725–739.

Carney, D., 1998. *Approaches to sustainable livelihoods for the rural poor*. O.D.I. Poverty Briefing, 2, January. London: Overseas Development Institute.

Chenery, H., *et al.*, 1974. *Redistribution with growth*. Oxford University Press.

Chilonda, P., Machethe, C., and Minde, I., 2007. *Poverty, food security and agricultural trends in Southern Africa*. ReSAKSS Working Paper 1. International Crops Research Institute for the Semiarid Tropics

(ICRISAT), International Food policy Research Institute (IFPRI) and International Water Management Institute (IWMI), Pretoria.

Cleaver, H., 1972. The contradictions of the green revolution. *The American economic review*, 62 (1/2), 177–186.

Cooke, B. and Kothari, U., eds., 2002. *Participation: the new Tyranny*. London: Zed Books.

Daniels, P., *et al.*, 2008. *An introduction to human geography: issues for the 21st century*. 3rd ed. Harlow: Pearson Prentice Hall.

Dorward, A., *et al.*, 2004. Institutions and policies for pro-poor agricultural growth. *Development policy review*, 22 (6), 611–662.

Economic Commission for Africa, 2003. *Towards a Green Revolution in Africa: harnessing science and technology for sustainable modernisation of agriculture and rural transformation*, (SMART/AGRI). Addis Ababa.

Estudillo, J.P., 2006. The Green Revolution, development of labour markets and poverty reduction in rural Philippines 1985–2004. *Agricultural economics*, 35, 399–407.

FAO/World Bank, 2001. *Farming systems and poverty: improving farmers' livelihoods in a changing world*. Rome and Washington, DC: FAO and World Bank.

Freidmann, J., 1966. *Regional development policy: a case study of Venezuela*. Cambridge, MA: MIT Press.

Funnell, D., 1991. *Under the shadow of apartheid: agrarian transformation in Swaziland*. Aldershot: Avebury Press.

Gadgil, M. and Guha, R., 1995. *Ecology and equity: use and abuse of nature in contemporary India*. New York: Routledge.

George, S., 1976. *How the other half dies: the real reason for world hunger*. London: Penguin.

George, S., 2007. Reflections. *Development and change*, 38 (6), 1161–1171.

Gleneagles Agreement, 2005. Available from: www.g8.gov.uk/ [Accessed 21 January 2011].

Gommes, R., 1996. *Rainfall variability and drought in sub-Saharan Africa since 1960*. Rome: Food and Agriculture Organisation of the United Nations.

Govereh, J. and Jayne, T., 2003. Cash cropping and food crop productivity: synergies or trade-offs? *Agricultural economics*, 28 (1), 39–50.

Government of Swaziland, 1999. *Swaziland household income and expenditure survey*. Mbabane: Central Statistical Office.

Government of Swaziland, 2001. *Swaziland household income and expenditure survey*. Mbabane: Central Statistical Office.

Hall, R., 2011. The next Great Trek? South African commercial farmers move north. *International conference on global land grabbing*, 6–8 April 2011, Co-hosted by the Land Deal Politics Initiative, *Journal of peasant studies and future agricultures consortium*, The Institute of Development Studies, University of Brighton, Sussex.

Hirschmann, A.O., 1958. *The strategy of economic development*. New Haven, CT: Yale University Press.

IFAD, 2001. *Rural poverty report 2001: the challenge of ending rural poverty*. Oxford University Press for IFAD.

Irz, X., *et al.*, 2001. Agricultural productivity growth and poverty alleviation. *Development policy review*, 19 (4), 449–466.

Key, N. and Runsten, D., 1999. Contract farming, smallholders and rural development in Latin America: the organisation of agroprocessing firms and the scale of outgrower production. *World development*, 27 (2), 381–401.

Komati Downstream Development Project, 2001. Available from: www.ecs.co.sz/komati/project_financing.htm.ecs.co.sz/komati/project_financing.htm [Accessed 31 January 2009].

Lappe, F. and Collins, J., 1988. *World hunger: twelve myths*. New York: Grove Press.

Little, P.D. and Watts, M.J., eds., 1994. *Living under contract, contract farming and agrarian transformation in Sub-Saharan Africa*. Madison, WI: University of Wisconsin Press.

Longhurst, R., 1988. Cash crops, food security and nutrition. *IDS bulletin*, 19 (2), 28–36.

Niazi, T., 2004. Rural poverty and the green revolution: the lessons from Pakistan. *The journal of peasant studies*, 31 (2), 242–260.

Polak, P., 2005. Waterand the three revolutions needed to end rural poverty. *Water science and technology*, 51 (8), 133–143.

Porter, G. and Phillips-Howard, K., 1997. Contract farming in South Africa: a case study from KwaZulu Natal. *Geography*, 82 (1), 38–44.

Potter, R.B., *et al.*, 2004. *Geographies of development*. 2nd ed. Harlow: Pearson Prentice Hall.

Rouault, M. and Richards, Y., 2003. Intensity and spatial extension of drought in South Africa at different time scales. *Water SA*, 29 (4), 489–500.
Scoones, I., 1998, *Sustainable rural livelihoods: a framework for analysis*. Institute of Development Studies, Working Paper No. 72, Brighton.
Shiva, V., 2008. *Soil not oil: climate change, peak oil and food insecurity*. London: Zed Books.
Shrestha, N., 2008. Misery is my company now: Nepal's peasantry in the face of failed development. *The journal of peasant studies*, 35 (3), 452–475.
Stringfellow, R., 1996. *An investigation of the organizational features, commodities and situations associated with contract farming and outgrower schemes in sub-Saharan Africa and of the factors which are critical to their successful operation*. London: Natural Resources Institute for the Overseas Development Administration.
Terry, A., 1997. *The Komati Pilot Project: a baseline survey and monitoring and evaluation system*. Mbabane: Ministry of Agriculture and Co-operatives, Government of Swaziland.
Terry, A., 2001. Who benefits from rural development projects? A case study of the Komati Pilot Project, Swaziland. *South African geographical journal*, 83, 18–27.
Terry, A., 2007. The Komati downstream development project: achievements and challenges. *Tijdschrift voor economische en sociale geografie*, 98 (5), 641–651.
Terry, A. and Ryder, M., 2005. Coping with change: the transition from subsistence orientated rain-fed agriculture to commercial irrigated agriculture. *Geography*, 90 (2), 138–150.
UNICEF, 2006. Available from: www.unicef.org/infobycountry/swaziland.html.unicef.org/infobycountry/swaziland.html [Accessed 20 January 2009].
United Nations Development Programme, 2006. *Report*. New York: United Nations.
Van Waveren, E., 2003. *Land resource distribution under customary tenure in Swaziland: a geographic analysis with special attention to semi-arid land*. Unpublished PhD thesis. Faculty of the Built Environment, University of the West of England, Bristol.
Van Waveren, E., 2006. How compatible is customary tenure with the aims of the Swaziland environmental action plan? *In*: J. Hill, A. Terry and W. Woodland, eds. *Sustainable development: national aspirations, local implementation*. Aldershot: Ashgate, 205–227.
Van Waveren, E., Nhlenetfa, J.V., and Mashwama, J.S., 1994. *Land use planning for rational utilization of land and water resources*. Vols. 1–15. FAO, UNDP Project SWA, 98/0001, mimeographed, MOAC, Mbabane.
Whiteside, A. and Whalley, A., 2007. *Reviewing 'Emergencies' for Swaziland shifting the paradigm in a new era*. National Emergency Response Council on HIV/AIDS (NERCHA), Mbabane. Available from: www.data.unaids.org/pub/Report/2007/swaziland%20emergency%20report_final%20pdf_en.pdf [Accessed 10 January 2009].
World Bank, 2000. *Swaziland: reducing poverty through shared growth*. Washington, DC: World Bank.
World Bank, 2003. *SWAZILAND – energizing rural transformation*. Report No. AB421, Initial Project Information Document, Washington, DC.
World Bank, 2006a. *Promoting increased fertilizer use in Africa: lessons learned and good practice guidelines*. Africa: World Bank.
World Bank, 2006b. *Sustainable land management: challenges opportunities and tradeoffs. The agriculture and rural development department*. Washington, DC: World Bank.
World Bank, 2008. *World development report*. Washington, DC: World Bank.
Xinshe, D. and Headey, D., 2008. Toward a green revolution in Africa: what would it achieve, and what would it require? *Agricultural economics*, 39, 539–550.

Malawi's agricultural input subsidy: study of a Green Revolution-style strategy for food security

Marie Javdani

Department of Geography, University of Oregon, 7554 McLaren Ave., West Hills, CA 91307, USA

Development scholarship and policy have recently returned to endorsing a 'new' Green Revolution (NGR) in Africa for the promotion of food security and economic development. The attention received by such programmes as the Alliance for a Green Revolution in Africa and the Millennium Village Projects are some prominent examples of this. Meanwhile, the government of Malawi has been lauded internationally for having averted widespread hunger within its borders through a subsidy that provides discounted chemical fertilizers and improved seeds to smallholder farmers. Many view the increased use of chemical fertilizer in Malawi agriculture and the resultant rise in maize yields – described by such publications as the *New York Times* as the 'Malawi Miracle' – as evidence that the prescribed NGR is indeed a recipe for success. Based on household-level surveys from the Zomba District of southern Malawi, this paper argues that although this subsidy programme seems to have filled the national 'food gap' in Malawi, shortfalls continue at the household level because of unresolved production issues and household vulnerability. I argue that the problems of accessibility and effectiveness of the subsidy are symptomatic of the inequalities inherent in the current food system, as described in interviews with respondents. These weaknesses are characteristic of NGR-style food security strategies and while programmes can be adjusted, they will remain subject to the same systemic processes that allow some to prosper while others want.

In 2007, Malawi caught the attention of the international development community with its introduction of an agricultural input subsidy whose goal was to eliminate widespread hunger within its borders. It seemed Malawi was finally on its way to becoming food secure while at the same time setting an example for the rest of the developing world in need of better hunger policy. After spending the better part of 25 years obeying the dictates of lending agencies such as the World Bank and International Monetary Fund, Malawi decided to, as aptly described by Celia Dugger of the *New York Times*, 'follow what the West practices, not what it preaches' (Dugger 2007). In other words, rather than continuing to adhere to the dogma of free-market and counter-protectionist policy, Malawi has decided to help itself.

Malawi's Agricultural Input Subsidy Programme (ISP) provides chemical fertilizers and hybrid seeds to a targeted group of smallholder farmers whose productive potential was judged to be hampered by inadequate soil fertility. With the inception of the policy in 2005 came promises from the government that Malawi's need to go begging for food was in the past. Indeed, the following two agricultural seasons yielded record harvests. Analysis of data from the National Statistics Office (NSO) and independent research suggest that this increase is largely attributable to the subsidy (Denning *et al.* 2009, Dorward *et al.* 2010, Dorward and Chirwa 2011). The NSO reports that national annual yields have remained above the baseline nutritional requirement set by the Ministry for Agriculture and Food Security (MoAFS). However, development scholars have long argued that analysis using only gross national estimated crop yield overlooks considerable variations in production and distribution at sub-national levels. It is therefore crucial to examine food security at the community and household levels as well.

The story of Malawi's ISP seems to lend support to the recently revived enthusiasm for the promotion of agricultural technology within development circles (Sachs 2005, World Bank 2007, Denning *et al.* 2009). The apparent solution in a world where millions starve and billions suffer from malnutrition is to produce more food. Thus, proponents of the Green Revolution (GR) look back at its influence in Asia where miracle seeds and agricultural chemicals paved the way to food self-sufficiency. However, a look at patterns of hunger within Malawi and worldwide tells us that scarcity is only part of the problem. Indeed, despite the unquestionable benefits gained through the input subsidy, hunger remains a serious issue in Malawi.

Using the example of the ISP, this research asks why, despite many nations having filled the national 'food gap' through agricultural technology, do individual and local food security remain elusive? This study uses household-level data for subsidy receipt, harvest and duration of household maize storage to complement national-scale analyses by Dorward and Chirwa (2011) and Jayne (2010). I argue here that the ability to most effectively obtain and use agricultural technology, subsidized or otherwise, lies in the same pool of resources that allow some to ensure their own food security while others cannot. Moreover, evidence from interviews suggests that the problematic distribution of benefits of the subsidy reflects, if not reinforces, a food system that already disadvantages the poor. This finding supports Peters' (2006, p. 332) conclusion that '... common events and crises at the national level ... provide opportunities that benefit some but maintain or increase obstacles for others' and are thus experienced by different categories of the population in diverse ways.

This paper is not meant as a condemnation of agricultural input strategies, but rather as an argument that underlying inequalities prevent this type of programme from being more than merely a stop-gap. After setting the context of the GR in Africa and Malawian agricultural policy, I analyse three elements of the subsidy on a local level: distribution, harvest size and grain storage. A discussion of these factors in combination with other sources of household food security and well-being demonstrates a pattern of inequality preventing smallholders from making long-term improvements to their lives. Data in this paper represent a 'village-eye-view' snapshot in time from research conducted in the dry season of 2008.

Given the comparisons drawn between the ISP and programmes such as the United Nations Millennium Village Projects (MVPs) and the Alliance for a Green Revolution in Africa (AGRA) (Denning *et al.* 2009), it is important to understand the effects of the ISP and its lesson for the GR movement. I suggest here that while these programmes can be adjusted to increase their effectiveness, they will continue to underperform if the problematic underlying principles of the GR are not reconsidered.

Reconsidering the Green Revolution

The GR was introduced in the decades after World War II as a scheme to inject modern agricultural technologies into the struggling farming sectors of developing countries. The goal was to promote food security and economic development by raising agricultural production and opening economies to commodity markets. Scientists designed 'improved' varieties of certain staple crops, beginning with rice and wheat. The use of improved seeds, along with chemical fertilizers and pesticides, had striking results: the case most commonly cited is India, whose grain production increased threefold nationally, and as much as 10-fold in Punjab and other districts; meanwhile, many nations that had previously been required to import grain in order to fill the gap between national grain production and requirement became net-exporters.

The results of the GR are often touted as a guide for economic development in low-income countries. The original underlying principle was that to produce more food is to feed more people. As a measure of its success, proponents cite the marked rises in agricultural production and economic indicators such as gross domestic product (GDP), exports, wages and food price stability (Carr 1997, Collier 2007, Morris *et al.* 2007). The use of these indicators is based on the assumption of modernization theory that development can be judged by the degree to which a national economy is engaged with the global capitalist economy (Bryant and Bailey 1997). This argument relies on the ability of market forces to distribute goods based on consumption needs. However, dissenters argue that certain problems accompany the introduction of such technology. Often cited are negative ecological effects, farmer debt, land and wealth concentration, the failure of markets to evenly distribute the benefits of higher production rates and a growing disparity between rich and poor populations.

Not all arguments regarding the disadvantages of GR technology are dismissals of market and technological innovation entirely. Although some argue that despite its success in boosting food production, the GR has 'proved an environmental, social and human disaster' (Sesmou 1991, p. 47; see also Holt-Giménez *et al.* 2006, Daño 2007), others argue that the GR itself is not to blame, but rather a 'social order permitting a tight grip on resources by only a few' (Lappé and Collins 1998, p. 53). As the ISP has proven in Malawi, increasing production of staple crops through lower prices for inputs is a vital part of promoting rural food security. But the differential distribution of these benefits between richer and poorer households is a signal of the need to critically examine the role of the GR in the context of pre-existing global and local inequalities.

Meanwhile, a substantial portion of development resources continue to focus on technology to boost production and expand agricultural markets. For example, the objectives of the AGRA include the promotion of agricultural technology by educating African farmers on its benefits and supporting expanded markets for the distribution of both inputs and produce (Alliance for a Green Revolution in Africa 2011). AGRA's supporters tend to downplay or ignore problems created or neglected by the GR in Asia, while also failing to question the underlying assumptions behind the push for technological innovation – for example, based on pre-conceived notions of a problem (and therefore its solutions) with little consideration of local contexts (Carr 2008). Most frequently cited is the notion of a 'low-productivity trap' in which low rates of production require the majority of household resources to be devoted to agriculture simply in order to survive (Carr 1997, Sachs 2005, Collier 2007, Morris *et al.* 2007, World Bank 2007, Denning *et al.* 2009, Dorward and Chirwa 2011). Although it is very much the case that households that spend a large proportion of their resources on food have little opportunity to improve their lives, the 'productivity trap' argument is problematic, as proponents see low productivity as the causal factor, diagnosing the problem as a lack of production technology. Rather, this paper argues that vast socio-economic inequalities and distribution of resources (and power over resources) are the underlying causes of low production and its accompanying consequences. This is a problem not simply amenable to injections of agricultural technology.

Agricultural development in Malawi

Malawi is a deeply impoverished sub-Saharan African country with an estimated population of 14 million. The country is primarily dependent on rain-fed agriculture, with one rainy season per year. Maize, grown almost universally by Malawian smallholders, represents the bulk of agricultural production and dietary intake. Maize production is primarily for in-home consumption, whereas export crops include burley tobacco (the current leading export, at 53%), tea, sugar, cotton, coffee and peanuts. However, Peters (2006) highlights the mistaken tendency of assuming that maize is the single crop upon which Malawian food security depends. Although maize is a vital component of the Malawian diet, it is important to note that many crops, including rice, cassava, millet and various fruits and vegetables, are consumed throughout the year, and are often eaten as a supplement to dwindling maize reserves.

The first few decades of the post-independence era (Malawi became independent in 1964) saw little change from pre-colonial trends. White-owned estates were annexed, some of which were broken up for smallholders, whereas many were preserved for the African elite. Agriculture Development and Marketing Corporation (ADMARC) continued to serve the interests of estate owners, and commodity production remained the central concern of agricultural policy (Harrigan 2001). With the arrival of the debt crisis of the 1980s came International Monetary Fund (IMF)-mandated Structural Adjustment Programmes, resulting in the reduction of government services, a dismantling of ADMARC and the liberalization of burley tobacco markets to allow non-estate farmers to participate in its production and sale. Harrigan (2003) describes an ongoing tension between the Malawian Government, whose priority was boosting agricultural productivity and the international aid community, who argued that the best way to promote food security is through price controls and minimal state involvement. She finds that despite two decades of attempted cooperation this tension was never resolved, thus 'preventing a consistent strategy for the agricultural sector' (p. 848). In the Zomba study area, Peters' (1996) conclusion was that although the income-producing opportunities of market liberalization benefited the wealthier households, the political–economic situation of the poorest resulted in a relative worsening of well-being in terms of food security and income. In fact, throughout her two decades of research, she continued to observe a pattern in which changes in economic and agricultural policy disproportionately favoured the better off (see Peters 1989, 1992, 1996, 2006).

Between IMF restructuring schemes, the government of Malawi experimented with a series of 'pro-poor' food security strategies. Notable was the Starter Pack (SP) programme in 1998 and 1999, later scaled back to a Targeted Input Programme (TIP) in 2000. The SP provided small amounts of fertilizer and seeds (enough for 0.1 ha), along with agricultural extension, to all smallholders. The results were encouraging: in 1999, SP provided an average of 175 kg more maize per household, reducing the hunger season by two to four months; the Ministry of Agriculture reported a national surplus of 0.5 metric tonne (Levy and Barahoma 2002). Although SP was meant to be universal, TIP targeted the poorest smallholders. Input packs were to be distributed to the most 'deserving' households selected by a village committee. However, this strategy lacked transparency, which, combined with criteria that insufficiently identified proper beneficiaries, resulted in problematic distribution, as many better-off households received packs while many poor households did not (Levy and Barahoma 2002, Chinsinga 2005).

Input subsidy programme

The residing government of President Bingu wa Mutharika implemented the current ISP starting in the 2005/2006 crop season. It was designed to provide reduced-priced agricultural inputs to 'key' producers in the smallholder sector. Owing to vague guidelines for selecting recipient households – an unresolved problem of the TIP programme earlier in the decade, targeting of the ISP has been problematic from the start. However, positive results in local and national production have been substantial. The government of Malawi has reported a grain surplus every year since ISP's inception.[1]

Popular media have widely reported the positive results of the ISP. Referred to as the 'Malawi Miracle', the ISP is often said to be a 'model [of development] for all of Africa'. Meanwhile, development and non-profit organizations have not been deaf to this praise. AGRA is one example. Another notable example is the United Nations' MVP headed by economist Jeffrey Sachs (see Sanchez *et al.* 2007). The objective of the MVP is to provide a baseline of public-sector goods in order to encourage private-sector savings and investment (Sanchez *et al.* 2007). Denning cites ISP achievements in fertilizer use and maize productivity as parallel to MVP goals, which focus on 'stimulating agricultural productivity [which will] likely increase commercial activity ... and extend new opportunities for agricultural input suppliers' (Denning *et al.* 2009, p. 6). Although somewhat of a departure from the free-market enthusiasm of recent decades, these objectives remain market- and technology-focused.

Dorward and Chirwa (2011) report many successes of the ISP, most notably increased smallholder use of inputs and higher rates of production. According to Dorward and Chirwa, 60–90% of fertilizer sales were attributable to the subsidy from the 2005/2006 season to the 2008/2009 season, whereas the use of subsidized fertilizer was estimated to account for 38–55% of increased national maize production. According to Dorward and Chirwa, the direct impacts of the subsidy were reduced poverty and enhanced food security as a function of increased average household harvests, stable food prices, higher wages for agricultural labour and growth in incremental profitability from input investment. The authors also observed higher GDP growth and development of improved input supply systems as secondary benefits (2011).

In another study, Ricker-Gilbert and Jayne (2011) found that consistent receipt of subsidized fertilizer over a three-year period resulted in compounded gains in production from one year to the next. However, they also found that although other indicators of household well-being such as household assets were often more favourable in years of receipt, this benefit was observed only during years of receipt and did not continue into the next year. That is to say, the benefits of bigger harvests, stable food prices and higher wages that accompanied the subsidy were impermanent, affecting most households for the current agricultural season only. This finding brings into question the claim by some advocates that enduring rural poverty is a result of being caught in a 'low-productivity' trap (Carr 1997, Sachs 2005, Collier 2007, Denning *et al.* 2009, Dorward and Chirwa 2011). Ricker-Gilbert and Jayne's finding suggest that boosting production, even for a period of years, does not result in a general 'breaking-free' of this trap.

Dorward and Chirwa (2011) also note some significant barriers. Most notably, they cite a failure of the coupon distribution system to ensure more direct impact of the subsidy to the poor. Some reasons for this include vague guidelines and lack of transparency in recipient selection, favouritism and patronism by distributors, and decreased rate of return for input use by the poor. Also significant, according to Dorward, Chirwa and Jayne (2010) were the increasing price of inputs and resultant over-budget spending, which detracts from services such as rural healthcare and other forms of hunger relief. The authors suggest public and more effective recipient selection, improved logistical and monitoring systems, complementary policies and investments, strong political commitment to the outcome of the subsidy and flexibility and innovation in programme implementation.

Methods

The case study for this paper is the Zomba District in southern Malawi. Population density and poverty rates here are among the highest nationwide. This results in small land holdings and thinly spread public and private resources. The data presented in this paper were collected in six village clusters south of Zomba town, neighbouring one of the United Nations' MVPs – an area known to some as a 'hunger hot-spot' (Denning *et al.* 2009). Villages are governed by an elected Village Head, who is the village representative to the Group Village Head or the Traditional Authority of the area.

This paper is based on fieldwork conducted during the dry season (roughly June–September) of 2008. The respondent pool was originally selected in 1985 at the beginning of a longitudinal study by Pauline Peters.[2] The original purpose of the Zomba study was to assess the role of burley tobacco cultivation and sales in smallholder livelihoods. For this reason, respondent households were selected such that 30% were households that cultivated and sold dark-fired tobacco. As the original population aged, Peters periodically expanded the number of households studied to include the households of the respondents' daughters. Since the production of burley tobacco relies on a minimum resource base, the study population is not economically representative of Malawi as a whole. The sample remains useful, however, because it represents a population that has the means of producing but struggles with a lack of agricultural resources, which is the stated target population of ISP distribution.

The respondent population consisted of 217 households, with a mean occupancy of 5.5 persons. Population age ranges from 0 to 85+ years, the majority of which (41%) are in the 0–15 years range. The population age structure suggests a high rate of HIV-related deaths in that the 25–45 age bracket has dropped from 22 to 13% of the study population in the 20 years since the Zomba study began in 1986 (Peters *et al.* 2008) – although because of sample size and bias these figures should be taken as suggestive only. HIV prevalence is currently estimated at 15% for Zomba District and 10.6% for Malawi as a whole. For the Zomba District, this figure has fallen by 3% since 2004 estimates. Peters *et al.* (2008) describe the respondent households as resilient in surviving this sharp decline in working-age adults, which they attribute to a robust ability of the extended family to mitigate shocks through resource and burden sharing.

Food security in the study area depends largely on a successful yearly crop season. The season begins with planting in November or December, with harvest taking place during May. It is rare for a smallholder household to produce enough maize to feed themselves until the next harvest. A 'hunger season' ensues once household maize stores have finished. Therefore, smallholders employ a number of alternative sources of food and income. These include hired agricultural and non-agricultural labour, cash-cropping, off-season production and market enterprises. This paper later argues that these strategies are likewise constrained by limited resources.

The 2008 study was a village-eye-view of the smallholder experience of the ISP. Two hundred and seventeen households were interviewed to determine the distribution of subsidized inputs and their benefits among different categories of the sample population. Detailed interviews were conducted at a subset of 30 households in three of the sample villages. While the size and manner of selection of the subject pool yield data that cannot be considered statistically significant, the information gathered can nonetheless be useful in a qualitative assessment of local-level results of the ISP.

Coupon distribution

The argument of this paper is not that poverty and inequality are the source of uneven results of the subsidy, but rather that these phenomena share a common cause in the tendency of accumulation to reinforce itself. What makes this study different from previous national-scale research is its focus on respondent experiences. To this end, this section provides an analysis of the quantitative data as a background to individual respondent perspectives of the subsidy. Hence, the links between variables observed here should not be taken as evidence of a causal relationship, but rather as a characterization of subsidy distribution and effectiveness as experienced by households in different indicator categories.

Although ISP programme documents describe the target population as 50% of smallholder households, particularly those seen as potentially productive but struggling (MoAFS 2007, p. 12), overall coupon receipt in the Zomba study was 84% (187 of the 217 total). Receipt was skewed slightly towards the upper expenditure and asset quartiles, with a lower percentage of households in quartiles one and two receiving coupons than the sample average. Although total coverage varied relatively slightly (89.7% of expenditure quartiles three and four vs. 81.7% of quartiles one and two receiving subsidized fertilizer), there was a substantial difference in the number of coupons per household. Recipient households in expenditure quartile one had an average of 1.2 coupons, whereas those in the fourth quartile averaged 2.1. The pattern was similar among asset quartiles with 1.1 and 2.0 coupons on average for quartiles one and four, respectively. Thus, while more households overall received some portion of the subsidy, a very few received the amount of fertilizer intended by the ISP. This dilution effect was more prominent in the lower quartiles.

Although the ISP document calls for each targeted household to receive two coupons, one for 50 kg of Nitrogen, Phosphurus and Potassium Complex as a basal dressing and the other for 50 kg of nitrogen-rich urea or Calcium Ammonium Nitrate ammonium nitrate as a top dressing, distributors (Village Heads and Group Village Heads) reported prioritizing extending coverage to a higher proportion of households over the number of coupons received per household. Out of 217 households, only 39 reported to have coupons to meet the criteria of the ISP. This number falls to 21 if households that purchased additional coupons are excluded.

The upper quartiles (both expenditure and asset) were buffered by their ability to purchase additional coupons or fertilizer at commercial prices. Of the 30 households that purchased coupons, 21 were in the top two expenditure or assets quartiles. The average land held by coupon purchasers was 1.59 ha (0.38 ha above the sample average of 1.21).

Respondents frequently indicated being dissatisfied with the distribution process. Interviewees unanimously agreed that 'the government' had failed to provide enough coupons. Twenty-four percent of respondents complained of some form of unfairness, favouritism, corruption or theft. Although there can be little doubt about some instances of abuse, ISP planning is largely to blame: distribution criteria are vague and difficult to interpret (e.g. 'vulnerable but hard-working'), which leads to the number of apparently 'deserving' households far outweighing the resources of the subsidy.

A comparison can be drawn between these results and Levy and Barahoma's (2002) assessment of TIP distribution, wherein they challenge the philosophy of TIP distribution as culturally inappropriate. They cite the problematic top-down assumption that scarce resources should be directed to the poor, which can lead to resentment and (real or perceived) favouritism, whereas in many villages cultural values 'dictate that (public assistance for growing food) should be *shared*' (Levy and Barahoma 2002, p. 13; *emphasis in original*). Chinsinga made similar observations from the 2002/2003 and 2003/2004 Winter TIP. He found that communities, rather than following MoAFS guidelines to direct inputs towards poor households with greatest potential for winter cultivation (i.e. possession of a *dimba* garden), distributed TIP resources according to locally derived criteria: perceptions of need, equity and entitlement (Chinsinga 2005, p. 286). In a South African study, Adato and Haddad (2002) found that some communities preferred random selection of members to targeted selection (cited in Chinsinga 2005).

Improving the distribution of coupons depends on a clear understanding of the purpose of the subsidy. Proponents who praise the subsidy's ability to improve national food security, stabilize local food prices and increase wages argue that directly targeting only the very poor reduces the subsidy's effectiveness. The distribution criteria in the ISP document targeting the most promising producers reflect this philosophy. Local priorities treated the subsidy more like a social supplement. When asked who should receive the subsidized inputs, most respondents named the poorest, households caring for the sick and for orphans, the elderly and households with very little land. Adjustments to the distribution process should better clarify intended recipients and allow for locally derived criteria.

Food security

Food security was assessed based on harvest size and duration of household maize stores. In 2006, Peters directly observed harvest and storage time for each household. During the 2008 study, respondents were asked to recall how long their harvests from 2007 had lasted and estimate when their 2008 harvest would run out. These data show that while poorer households benefited more from the subsidy as a proportion of their harvests in a year with favourable conditions, overall food security indicators remained much stronger in the better-off households. In addition, although on average the subsidy served as a buffer against poor weather during 2008, wealthier households experienced this effect much more strongly than the poor.

The 2006 harvest was a vast improvement over the disastrous one of 2005, whereas 2007 was even better. Peters recorded an average of 6.5 months of maize storage after the 2006 harvest, whereas respondents reported their 2007 harvest to have lasted an average of 9.25 months. However, harvest size was highly variable between households, some of whom found themselves searching for alternative sources of food very early in the agricultural cycle. Overall, indicators showed significant improvement in household food security over previous years, much of which is attributable to the increased yields resulting from the subsidized fertilizer, as well as favourable weather conditions.

The 2007/2008 growing season, however, was characterized by erratic, heavy rain and flooding, which disrupted the cultivation of maize and many other essential crops in the study area. Harvests suffered as a result, and respondents estimated on average that their food stores would last less than six months, a decrease of nearly three months from the previous year. This meant that the average respondent household would be searching for alternative means for more than half of the following year. Also, the number of total failures – that is, households harvesting less than one month's worth of maize – increased from a single household in 2007 to 32 households (21 of whom received coupons) in 2008.

Harvest response to coupon receipt was markedly positive. Compared with non-recipient households, the gross average increase of 119 kg (32%) harvested by recipients in 2008 was roughly uniform across all quartiles of both expenditures and assets. Thus, the poorer households experienced a much greater benefit as a percentage above their non-recipient peers – 78 versus 15% in the bottom and top quartiles, respectively. However, it remains that despite the benefits of the subsidy, non-recipient households in the top expenditure and asset quartiles harvested more maize than coupon recipients in the bottom three quartiles (see Figures 1 and 2). Harvest response to coupon receipt was most marked for households owning more than 1.5 ha of farmland. On average, recipients in this category nearly tripled the harvests of their non-recipient counterparts (837 vs. 247 kg).

In Figure 3 the strength of better-off households in benefiting from the subsidy is most evident. Among non-recipient households average harvest size varied positively with land ownership, but only by 90 kg across field size categories. Recipient households harvested 168, 150 and 305% of the average non-recipient in the first, second and third land categories, respectively.

In 2008, the subsidy provided a buffering effect against poor growing conditions, but disproportionately so for the better-off households. After a very successful growing season in 2007, household maize stores of both recipients and non-recipients lasted, on average, between 7.6 and 10.7 months across expenditure quartiles. Recipient households benefited by an average of 1.2 months over households that received no subsidized inputs. This pattern is observable

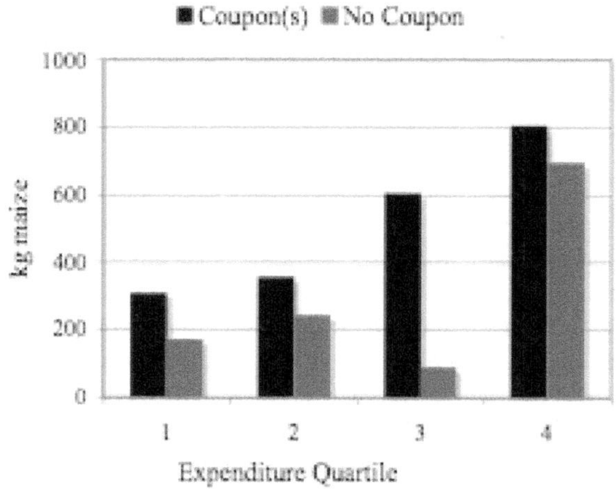

Figure 1. 2008 Maize harvest by expenditure.

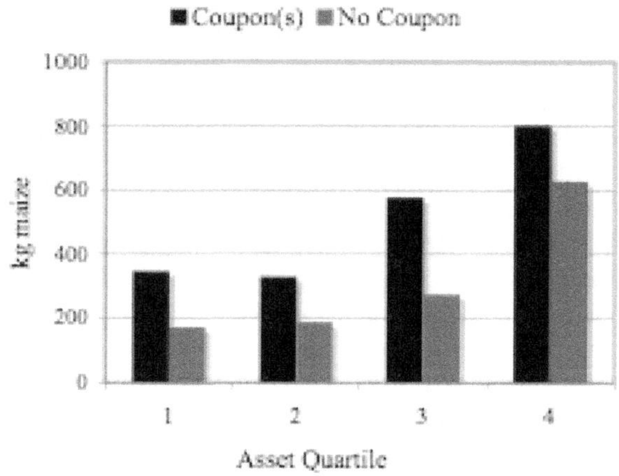

Figure 2. 2008 Maize harvest by asset quartile.

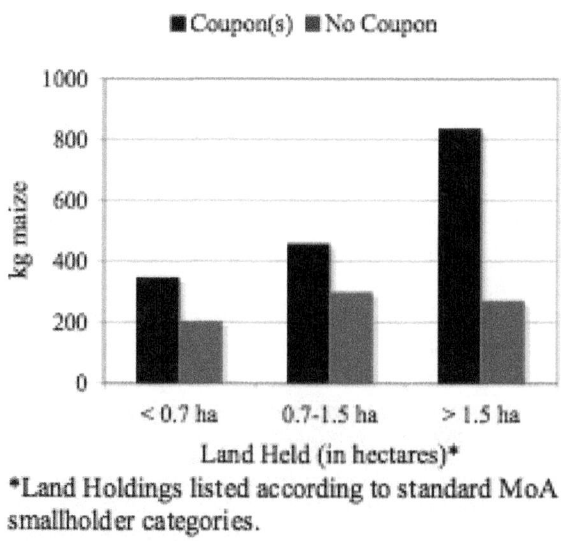

Figure 3. 2008 Maize harvest by land holdings.

across 2007 asset quartiles and land categories as well. In 2007, the maize harvests of recipient households in the top asset quartile lasted 6% longer than the harvests of their non-recipient peers – their stores having lasted 10.7 months vs. 10.1 months. However, in 2008 recipients in the top quartile experienced a substantial buffering effect of 3.4 months, (7.4 months compared with 4.0 months, a 45% difference). Meanwhile, the lowest asset quartile gained 1.8 months between recipients and non-recipients in 2007, but the difference in 2008 was less than one month. Thus, the buffering effect was much weaker for the poorer households during a critical crisis year (see Figures 4–7).

Household food security was never solely dependent on own-farm production. Many respondents who expected their own maize to last for only a short time commented that although this would cause them hardship, they were nonetheless unworried about food security. This section describes some individual accounts of the diverse resources that combine with their main-season maize harvest to meet household needs. Although the discussion is limited to a few key examples, all of the 2008 respondents reported utilizing some combination of different coping strategies.

After own-farm production, the most important source of livelihood for poor households was *ganyu*, casual day labour for other nearby farmers. Payment was generally in maize, although cash was increasingly common. Although some rely on *ganyu* only in times of severe need, it is more common for it to be a regular part of a household's resources (Whiteside 2000). Increased production provides more opportunities for *ganyu* (Dorward and Chirwa 2011) and gives labourers more bargaining power for negotiating wages (Peters 2006). For some, particularly those with little land of their own, *ganyu* can be the primary means of survival. Households with limited labour resources, for example female-headed households, those with high morbidity rates or young children to be attended, were less able to take advantage of labour opportunities – a problem similar to own-farm production. One respondent, a divorced woman from the second expenditure quartile and the lowest asset quartile, explained that in previous years production shortfalls were less of a concern than they were for her at present. Her mother had become chronically ill during the growing season, costing the labour of the sick woman as well as her caretaker. This limited the respondent's own production and her ability to perform *ganyu*. Although she received one subsidy coupon, which she attributes to her having harvested the two bags of maize she did manage, the respondent said that the fertilizer was less effective because she was unable to weed her gardens at key times during the growing season.

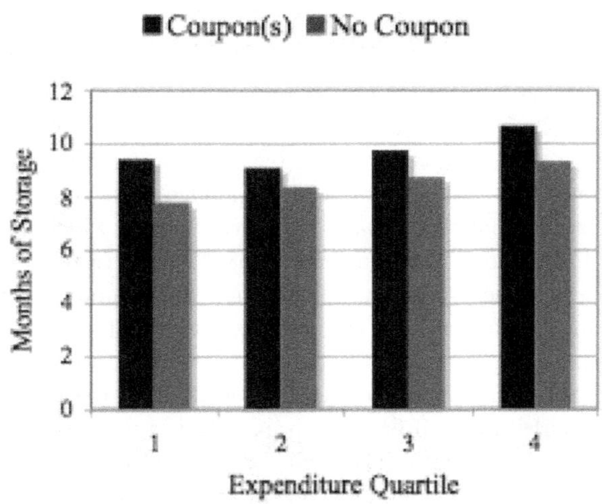

Figure 4. Months of maize storage by expenditure quartile, 2007.

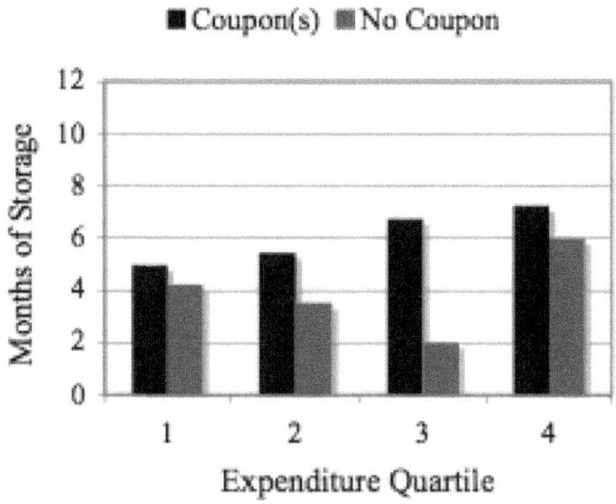

Figure 5. Months of maize storage by expenditure quartile, 2008.

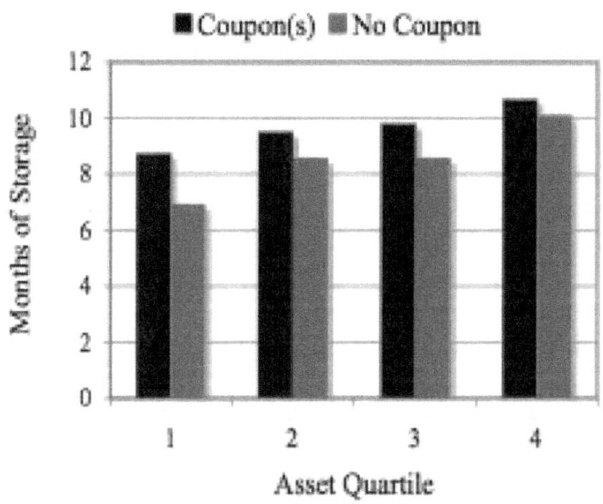

Figure 6. Months of maize storage by asset quartile, 2007.

River-adjacent gardens (*dimbas*) are also a supplemental source of food and income because they enable dry season farming. *Dimba* owners often grow off-season maize and vegetables to supplement their regular-season harvests, or use the gardens as a nursery for tobacco seedlings and other income-generating produce. Several respondents were to be found at the local markets selling vegetables from their *dimbas*, the income from which was a vital supplement to their diminished food stores after the poor 2008 harvest. However, *dimbas* are another example of a resource more accessible to the wealthier, landed households. Those respondents in the top land-owning category as well as the highest expenditure quartile were by far more likely to own a *dimba*.

Peters (2006) writes of the benefits to household welfare for those able to participate in the burley tobacco trade. Burley tobacco production exploded in the early 1990s with nearly 80% of the sample households participating. However, those whose production made a significant impact on household expenditures and assets were disproportionately those who already ranked in the upper quartiles, as well as those who owned sufficient land to support comparatively

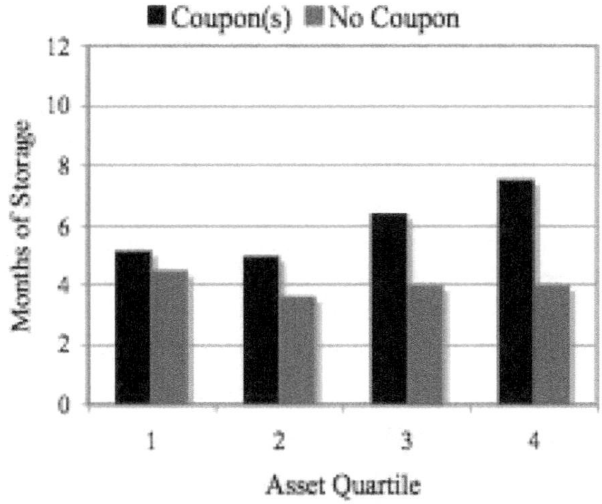

Figure 7. Months of maize storage by asset quartile, 2008.

larger-scale production. It was also these farmers who were more likely to be included in the farmers' clubs that negotiated the best prices for the product. By 2006, volatile international prices had resulted in the number of tobacco growers dropping to 65%. Respondents reported in 2008 reducing or eliminating burley tobacco production because of its rising costs and unreliable profitability. This was exacerbated by the poor maize harvest, after which farmers with the means decided to focus their dry-season production on supplemental maize and other foods, which they saw as lower-risk.

Household members might also find non-farm employment teaching, cooking or caretaking. Working-age males who left the household for employment in the cities or in the mining industry in neighbouring countries were the primary (although hardly the sole) demographic involved. However, extra household income opportunities were limited and generally low-paying. Higher-paid positions were more accessible to educated household members or those with particular professional credentials – a driver's license, for example. Remittances and gifts from these and other family members were important buffers against household resource constraints.

Conclusion

One could certainly argue that the subsidy has at least partially succeeded in its immediate objectives – increased fertilizer use and national and household maize production – but not in the ways attributed to it by popular media and high-level politicians. Equally, the expectation of GR advocates that local-scale market improvements and national-scale yield increases lead to sustainable progress towards individual and national self-sufficiency appears to be overly optimistic. The terrible harvest of 2008 was caused by factors well beyond the realm of agricultural technology: failed harvest due to heavy rains, global and local food price speculation and falling prices for high-value export crops. Thus, in a good year the subsidy enables household and commercial agricultural enterprise, whereas in a bad year it can provide a buffer against harvest loss – although disproportionately in favour of the better-off.

The observed relationship between indicators of household well-being and subsidy access and effectiveness ought not to be taken as conclusive evidence of a causal relationship. Here the merit of the low-productivity trap argument is important. Proponents of production for development argue that a household's inability to produce or procure sufficient food supply prevents its escape. The argument of this paper is not to deny the inherent feedback loops of poverty. Rather, the aim is to draw attention to the problems caused (or at least, not helped) by the disproportionate priority given to market-driven production strategies. Malawi took a step in the right direction when the government committed to providing more smallholders with one means of survival. Unfortunately, local, national and international politics fall short, missing the limits placed on production schemes by the disadvantages of the very poor. AGRA and projects like it are on the path to this same shortcoming. The faith of GR proponents in markets to 'find a way to bring [the food they need]' to the rural poor once low production and low income are overcome (Polak and Fry 2009, p. 177) neglects the differential command over the market enjoyed by the wealthy. Thus, the most important – and most frequently overlooked – part of the cycle of poverty is not just poverty reinforcing itself, but also of the growing power gap between the haves and the have-nots.

Acknowledgements

The author sincerely thanks Peter Walker and Pauline Peters, along with editors Bruce Scholten and Pratyusha Basu. Thanks are due to Daimon Kambewa, Noel Mbuluma and Dorothy Chirwa in Malawi. Thanks are also due to three anonymous reviewers and to Pauline Peters and Judith Carney for their valuable and thoughtful comments on previous drafts of this paper. And of course, the author thanks the Zomba study respondents. Any remaining weaknesses of this paper remain the author's own.

Notes

1. However, a recent study postulates that for political reasons, national yield figures may have been exaggerated by as much as 25% (Jayne 2010).
2. Pauline Peters (Harvard Institute for International Development and John F. Kennedy School of Government) began her longitudinal study in 1985, collecting data in 10–12 monthly rounds in 1986/1987, 1990/1991, 1997 and 2006, with short visits between these years. She was joined by Peter Walker (University of Oregon Department of Geography) and Daimon Kambewa (Bunda College of Agriculture, University of Malawi – Lilongwe) in later years. These studies provide background information for the research conducted by myself in the study area during the dry season (June–September) of 2008. Data from 2006 and earlier are Peters' and used with permission. Unless otherwise stated, all interviews and data for the time period after 2006 were part of the most recent (2008) stage of fieldwork.

References

Adato, M. and Haddad, L., 2002. Targeting poverty through community based-programs: experiences from South Africa. *Journal of Development Studies*, 38 (3), 1–36.

Alliance for a Green Revolution in Africa (AGRA), 2011. *Strategy for an African Green Revolution*. Nairobi: AGRA, Available from: www.agra-alliance.org [Accessed 21 June 2011].

Bryant, R. and Bailey, S., 1997. *Third world political ecology*. London: Routledge.

Carr, E., 2008. The millennium village project and African development: problems and potentials. *Progress in Development Studies*, 8 (4), 333–344.

Carr, S., 1997. A Green Revolution frustrated: lessons from the Malawi experience. *African Crop Science Journal*, 5, 93–98.

Chinsinga, B., 2005. A clash of voices: community-based targeting of safety-net interventions in Malawi. *Social Policy and Administration*, 39 (3), 284–301.

Collier, P., 2007. *The bottom billion: why the poorest countries are failing and what can be done about it*. New York: Oxford.

Daño, E., 2007. *Unmasking the new Green Revolution in Africa: motives, players and dynamics*. Penang: Third World Network.

Denning, G., *et al.*, 2009. Input subsidies to improve smallholder maize productivity in Malawi: toward an African Green Revolution. *PLoS Biology*, 7 (1), e1000023.

Dorward, A. and Chirwa, E., 2011. The Malawi agricultural input subsidy program: 2005–6 to 2008–9. *International Journal of Agricultural Sustainability*, 9 (1), 232–247.

Dorward, A., Chirwa, E., and Jayne, T.S., 2010. *The Malawi agricultural input subsidy programme, 2005/6 to 2008/9*. Lilongwe, Malawi: The World Bank.

Dugger, C., 2007. Ending hunger, simply by ignoring the experts. *The New York Times*, printed 2 December.

Harrigan, J., 2001. *From dictatorship to democracy: economic policy in Malawi 1964–2000*. Burlington, VT: Ashgate Publishing Company.

Harrigan, J., 2003. U-turns and full circles: two decades of agricultural reform in Malawi 1981–2000. *World Development*, 31 (5), 847–863.

Holt-Giménez, E., Altieri, M., and Rosset, P., 2006. *Food First Policy Brief No 12: ten reasons why the Rockefeller and the Bill and Melinda Gates Foundations' alliance for another Green Revolution will not solve the problems of poverty and hunger in sub-Saharan Africa*. Available from: http://www.foodfirst.org/en/node/1527 [Accessed 4 January 2011].

Jayne, T., 2010. *The 2008/09 food price and food situation in Eastern and Southern Africa: implications for immediate and longer run responses*. An International Development Working Paper. East Lansing: Michigan State University.

Lappé, F.M. and Collins, J., 1998. *World hunger: 12 myths*. New York: Grove Press.

Levy, S. and Barahoma, C., 2002. *2001–2002 Targeted input programme, main report of the evaluation programme*. Lilongwe, Malawi: Government of Malawi Ministry for Agriculture and Food Security.

Ministry of Agriculture and Food Security, 2007. *The 2007/2008 agricultural input subsidy programme*. Lilongwe, Malawi: Ministry of Agriculture and Food Security, Government of Malawi.

Morris, M., *et al.*, 2007. *Fertiliser use in African agriculture: lessons learned and good practice guidelines*. Washington, DC: The World Bank.

Peters, P.E., 1989. *Cash cropping, food security and nutrition: the effects of agricultural commercialization among smallholders in Malawi*. Cambridge, MA: Harvard Institute for International Development.

Peters, P.E., 1992. *Monitoring the effects of grain market liberalization on the income, food security and nutrition of rural households in Zomba South, Malawi*. Cambridge, MA: Harvard Institute for International Development.

Peters, P.E., 1996. *Failed magic or social context? Market liberalization and the rural poor in Malawi*. Cambridge, MA: Harvard Institute for International Development, Development Discussion Paper No. 562.

Peters, P.E., 2006. Rural income and poverty in a time of radical change in Malawi. *Journal of Development Studies*, 42 (2), 322–345.

Peters, P.E., Walker, P.A., and Kambewa, D., 2008. *The effects of increasing rates of HIV/AIDS-related illness and deaths on rural families in the Zomba District, Malawi: a longitudinal study*. RENEWAL Program.

Polak, P. and Fry, S., 2009. Water and the twin challenge of feeding 3 billion new people and ending rural poverty. *In*: M. Llamas, L. Martínez-Cortina and A. Mukherji, eds. *Water ethics*. London: Taylor & Francis Group, 171–203.

Ricker-Gilbert, J. and Jayne, T.S., 2011. *What are the enduring effects of fertiliser subsidy programs on recipient farm households? Evidence from Malawi*. East Lansing, MI: Department of Agricultural, Food and Resource Economics, Staff Paper 2011-09.

Sachs, J., 2005. *The end of poverty: economic possibilities for our time*. New York: Penguin Books.

Sanchez, P., *et al.*, 2007. The African millennium villages. *Proceedings of the National Academy of Science*, 104, 16775–16780.

Sesmou, K., 1991. The food and agriculture organization of the United Nations: an insider's view. *The Ecologist*, 21, 47–56.

Whiteside, M., 2000. *Ganyu labor in Malawi and its implications for livelihood security interventions: an analysis of recent literature and implication for poverty alleviation*. Blantyre: Oxfam International Programme in Malawi.

World Bank, 2007. *World development report 2008: agriculture for development*. Washington, DC: World Bank.

Challenges for under-utilized crops illustrated by ricebean (*Vigna umbellata*) in India and Nepal

Peter Andersen

Department of Geography, University of Bergen, Norway

Several research initiatives have promoted the revival of 'under-utilized' or 'orphan' crops, discursively portrayed as benign, pro-poor, adapted to marginal conditions, nutritious and modest in input requirements and strengthening agricultural sustainability. This paper presents the case of the under-utilized crop ricebean (*Vigna umbellata*), based on experience from the project FOSRIN (Food Security through Ricebean Research in India and Nepal). The crop is widely scattered in South Asia, but production is declining. This decrease is explained by obstacles throughout the value chains at a global scale (trade regulations, practice of Intellectual Property Rights-related rules for germplasm exchange) as well as at a local scale. The dis-adoption of the crop by farmers is linked to traits of existing landraces: growth habits, appearance and taste, and pest problems. This dis-adoption presents dilemmas for the biodiversity argument. If ricebean is to regain its position, efforts are needed in marketing and promotion and also in plant breeding and provision of improved seed material. It will require varieties that are able to meet farmers' desires in terms of growth habits, time to maturity, uniform seeds, good yield, resistance to pests and diseases, as well as able to fetch good market prices through attractive appearance, cooking qualities and organoleptic qualities. This will, inevitably, be at the expense of existing varieties with less competitive features, making the responsibility for conservation an issue for off-farm strategies.

Introduction

Among criticisms of the Green Revolution is the process of 'genetic erosion', which refers to the reduction of genetic variation within major crops and the loss of other crops and is assumed to lead to an increased risk of crop failure and malnutrition. The perception of genetic erosion as a threat is a major issue in public debates on agricultural modernization, as in, for instance, Shiva (1993). Intuitively, the replacement of a variety of landraces of traditional crops with improved varieties of a few, major crops will carry the risk of genetic erosion. However, agro-biodiversity dynamics is a complex field with many unknowns when it comes to firm data. While India's report on agro-biodiversity to FAO (NBPGR 2007, p. 17) states that 'Many of the landraces and primitive cultivars have already vanished and some are on the verge of it due to their abandonment by farmers in lieu of high yielding varieties. The remaining ones are genetically deteriorating gradually due to hybridization, selection or genetic drift', the global report (FAO 2010, p. 16) is more cautiously noting that 'it seems that general statements purporting to quantify the overall amount of genetic erosion that has occurred over the past decade are not warranted'.

This paper is addressing a special section of the global menu of biodiversity, namely crops that at various occasions are referred to as under-utilized, under-exploited, underdeveloped, neglected, orphan, indigenous, minor, native (or even tribal) crops or species. The acronym NUS (neglected and underutilized crop species) has been applied by the FAO (Padulosi *et al.* 1999). Underutilized crops are commonly linked to food security and nutrition (e.g. Fassil *et al.* 2000, Johns and Sthapit 2004), and especially to 'hidden hunger' due to the high content of micronutrients in several underutilized crops, and they have been connected to strategies to achieve the Millennium Development Goals (Ravi *et al.* 2006). As examples of the nutritional aspects, quinoa, amaranth and millets are commonly mentioned as species that are exceptionally rich in protein and minerals. Much research is still needed, however, to document the assumed nutritional benefits, at both the species and variety levels (Toledo and Burlingame 2006), in addition to their contribution in various diets. The underutilized crops are primarily grown by marginal farmers whose main adaptation to environmental uncertainties is risk spreading strategies through the application of mixed farming and multi-cropping systems with a high degree of agro-biodiversity. In addition to their food value, these crops will often play complex roles as integral parts of nutrient and energy flows within the farming systems.

Underutilized crops are almost invariably presented in connection with participatory, multidisciplinary stakeholder/partnership approaches to agricultural development (Johns 2004, Johns and Sthapit 2004, Padulosi and Hoeschle-Zeledon 2004) and in this respect belong to post-Green Revolution or Secondary Green Revolution development strategies. Framed and promoted with social, cultural, ecological and nutritional benefits, the under-utilized crops may, however, also be at risk of becoming romanticized as 'noble savages' among crops. If they are so good, why are they not spreading on their own? Will they inevitably lose out in the course of commercialization of agriculture and globalization of food trade?

The present paper is based on the case of ricebean (*Vigna umbellata*) and findings in connection with the EU Framework-6 project FOSRIN[1] – Food Security through Ricebean Research in India and Nepal. This project is aimed at promoting and developing ricebean as a means of supporting food security among marginal farmers, through breeding, development activities and basic research in a number of outreach sites in Western and Eastern Middle Hills in Nepal, and Himachal Pradesh and Assam in India. Ricebean carries a number of the beneficial traits mentioned above. However, the FOSRIN programme has found that the general trend in the traditional core areas in the Himalayan region is one of dis-adoption and decline. How can this be explained in the light of ricebean's ability to promote environmental sustainability by averting erosion, nurturing community sustainability with its nutritive benefits and fostering economic sustainability in times of high oil prices due to its low input requirements? The paper tries to address these questions both at a value chain level and at the level of farmers' decision making.

Sources and methods

This study draws on general experiences from the FOSRIN project, existing literature on ricebean as well as fieldwork in ricebean growing areas. In November 2005, I conducted a 'rapid rural appraisal' in the affluent and dynamic commercial agricultural district of Ilam in Eastern Nepal. The RRA was based on interviews with farmers growing ricebean and focused on farming practices, local uses and marketing aspects. In addition, the project included brief field visits in 2008 in Karbi Anglong District, Assam, and in 2009 in Himachal Pradesh, India.

As a comparative case, I present findings from a study by Acharya (2008) who conducted fieldwork in Dang District in Western Nepal over a period of about two months from May to July 1997. Acharya's research is based on semi-structured interviews and conversations with farmers and is important for understanding the reasons for dis-adoption of ricebean in a less developed region of Nepal. Thus, I am trying to bring together field evidence from a number of sites in a larger transect of the Middle Hills of the Himalayas.

Most of the existing scientific literature on ricebean is technical, relating to biological, agronomic or nutritional issues of ricebean. While following an objectivist tradition, many of the authors have written with the aim of evaluating and promoting ricebean as a crop with significant potential for human consumption. Technical reports that have been produced by the FOSRIN project (for instance Gautam *et al.* 2007, Khanal *et al.* 2007), which were based on inputs from partners of the project and used a variety of field methods, have also been consulted.

Defining and framing under-utilized crops

A definition of under-utilized crops is not quite straightforward. Padulosi and Hoeschle-Zeledon (2004, p. 5) suggest that the concept(s) commonly can be applied to 'those species whose potential has not been fully realized'. This however is too vague, and a further refinement of the definition is suggested to be: 'those non-commodity crops, which are part of a larger biodiversity portfolio, once more popular and today neglected by users' groups for a variety of agronomic, genetic, economic, social and cultural factors' (Padulosi and Hoeschle-Zeledon 2004). This definition contains interesting elements, since it also allows for factors that explain why farmers have decided not to cultivate it.

The crops that by various authors and agents are assigned to these concepts include:

- semi-grains and grains such as buckwheat (*Fagopyrum* spp.), finger millet (*Eleusine coracana*), quinoa (*Chenopodium quinoa*), amaranth (*Amaranthus caudatus*) and other Andean grains;
- roots and tubers such as achira (*Canna edulis*), maca (*Lepidium meyenii*) and yacon (*Smallanthus sonchifolius*);
- fruits such as noni fruit (*Morinda citrifolia*);
- vegetables such as African eggplant (*Solanum aethiopicum*);
- numerous lesser known pulses such as bambara groundnut (*Vigna subterranea*) and grass pea (*Lathyrus sativus*).

Researchers engaged with under-utilized crops include figures at some of the global institutions connected to the Consultative Group of International Agricultural Research (CGIAR) system and the International Board for Plant Genetic Resources, now called the International Plant Genetics Resources Institute. Another initiative, the Global Facilitation Unit for Underutilized Species, has been joined with the International Centre for Underutilised Crops to form the Crops for the Future, which is hosted by Bioversity International in Malaysia. Bioversity International was launched in January 2008 when 54 countries signed an establishment agreement. In addition, various FAO bodies have been involved in research on under-utilized crops, and there are likely to be large untapped synergies due to the spread over a large number of organizations. Some work in progress that could be mentioned is the merging of and refinement of internet sites, such as the Platform for Agrobiodiversity Research, http://agrobiodiversityplatform.org. Although the funds probably are only a fraction of what is spent on major, commercial crops, research on and promotion of under-utilized crops is receiving support from a number of national and international development agencies and foundations, and large and small NGOs. To summarize, the field of underutilized crop research is recognized and well-profiled, positively linked to the concept of biodiversity and self-evidently a tool against genetic erosion.

Under-utilized crops are generally presented on one side of a binary reflecting a conventional/ alternative division, where 'alternative' implicitly is assumed to be benign. In that respect they play an important role in anti-modernist discourses on agro-biodiversity such as those presented by the Indian activist Vandana Shiva (1993) who repeatedly emphasizes the benefits of mineral and nutrient content of under-utilized crops, and their prudence and efficiency in terms of water.

Padulosi *et al.* (1999) noted three major areas of interest and concern with respect to under-utilized crops that affect aspects of agricultural sustainability: (i) new markets and uses, (ii) environmental concerns and (iii) food security and nutrition. New markets and new uses are enabled by technologies and rising needs and opportunities. While one trend in the global food supply is towards homogenization and standardization of foods, the increasing consumer awareness in many markets represents a counter-trend that can create options for under-utilized crops locally and globally. New markets for biofuels offer a range of opportunities for some under-utilized crops, but also give some new dilemmas concerning food versus non-food production. Awareness of environmental change, land degradation and ecosystem stability has led to growing interest in crops that are adapted to climatic stress, marginal environments and few agricultural inputs. Genetic diversity is itself assumed to lead to higher ecosystem stability and lower vulnerability to risks of pests and diseases.

The increasing dominance of major crops has led to a decrease in the nutrient density of the diet of large population groups worldwide (Bouis *et al.* 2000). Within the CGIAR system, this has led to increasing focus on holistic views on food policies. This is done through increasing research emphasis on dietary diversity as well as adequacy rather than mere energy sufficiency. The Future Harvest programme led by the International Food Policy Research Institute focuses mainly on *biofortification* through improving the nutrient content of staple foods, based on the assumption that improved staples will reach some of the most marginal groups effectively. Some biofortification programmes on other crops than staples are also undertaken in the CGIAR system. A few of the Future Harvest projects use Genetic Modification technologies, and increasingly, the biotech companies argue for their products in terms of nutritional value.

The anti-modernist discourse, as presented by Vandana Shiva, rejects this logic and presents under-utilized crops as the alternative:

> ... all these traits already exist in farmers' varieties and farmers' fields. Legumes and pulses intercropped with cereals fix nitrogen (...) We do not need genetic engineering to give us crops rich in nutrition. Amaranth has nine times more calcium than wheat and 40 times more calcium than rice. (...) Nutritious and resource prudent crops such as millets and legumes are the best path of food security (Shiva 2000, p. 113).

Obstacles in value chain for under-utilized crops

The main problem for under-utilized crops is obviously that they have to compete with non-orphan crops, nurtured by the bulk productivity-oriented innovations of the Green Revolution, which is heavily biased towards a few grain and oil crops. These crops are dominant because their genetic potential originally was higher or more easily tapped than the potential of other crops, and their development has been supported throughout agricultural history. Both the private and the public sector have focused on the crops with the highest yield or profit potential first. On the other hand, this entails that the marginal benefit from development of under-utilized crops now may be higher than that of the major crops since the genetic potentials have not yet been exhausted. In a global trade perspective, export subsidies reinforce the productivity advantages of dominant crops. Thus, with respect to the international pulse trade, the notion of 'soy imperialism' (Shiva 2000) is not completely unjustified.

Although the CGIAR system has concern for under-utilized crops, it still prioritizes the so-called mandate crops, the species for which the research centres under CGIAR have specific development programmes. Currently, CGIAR is in the second phase of the Generation Challenge Programme, which aims at 18 mandate crops. Among the grain crops, rice, wheat, maize, barley, sorghum and millet belong to this group. Mandate pulse crops are chickpea, cowpea, Phaseolus beans, lentils, pigeon pea and soybean. The under-utilized crops remain a niche activity which is governed by Bioversity International, but without specific crop development programmes.

A common criticism of the Green Revolution, particularly during the first decades of implementation, was that it favoured production in the relatively uniform plains and valleys offering little for the complex ecological conditions faced by hill farmers. In Nepal, the spread of high-yielding varieties has primarily been in the Terai plains, while landraces and traditional crops have been predominant in the hill areas. This does not, however, imply that hill agriculture has been static. The existence of local 'landraces' of Latin American crops such as maize and potato in the Himalayan hills is clear evidence of how the farmers over generations have been innovators experimenting with introduced crops and varieties. Current developments in hill agriculture include some diffusion of high-yielding varieties, and commercial production of high-value crops such as vegetables, spices and fruit orchards. These changing cropping patterns do potentially compete with traditional crops, including the under-utilized. The speed and degree of this agricultural transition in the Himalayas is not well documented because it is gradual and appears pocketwise, depending on a complex set of factors such as road development, diffusion of innovation and technical knowledge and formation of trade networks.

However, there are significant impediments for the producers to reach international markets. Market regulations such as the EU Novel Foods Directive and the Novel Foods Regulation (EU 258/97) are at the outset meant to protect the consumer (Hermann 2009). This regulation 'arbitrarily defines novel food as food or food ingredients that were not used for human consumption to a significant degree within the EU before 15 May 1997' (Hermann 2009, p. 501), effectively preventing many crops from access to potentially profitable markets. The FAO and WHO *Codex Alimentarius* can also serve as an obstacle to market access for under-utilized crops (Chishakwe 2008).

The rules restricting the export of germplasm are argued in terms of protecting Intellectual Property Rights (IPR) of local crops on behalf of developing nations. However, they may in reality serve as an obstacle to international research and development of under-utilized crops which is needed for their survival. The FOSRIN project is specifically aimed at providing improved seeds as a public good, not least for poor and marginal farmers, but the Indian practice of regulations of germplasm exchange has created a particularly difficult situation for the project. After nearly four years of high-level applications, the project partners have not been permitted to bring ricebean germplasm out of India.

Other problems are linked to market access (Gruère *et al.* 2006): neglect from local traders and wholesalers in producing countries as well as the global supermarket chains and their tendency to strip down diversity. Finally, at the very local level, the rules for share cropping in Nepal present an example of another barrier to production of under-utilized species, since the landlord has the right to decide the choice of crops to be planted (Acharya 2008).

The case of ricebean

Ricebean (*V. umbellata*) is a multipurpose legume crop which is native to Indo-China. It carries numerous local names, in Nepal for instance *Marsyam*, *Jhilinge*, *Siltung* or *Gurans* in addition to dozens of varietal names. In Hindi, it is referred to as *sutari*, but many more local names are found in India, such as *Rajmoong*, *Navrangi*, *Satrangi*, *Moth* (not to be confused with *Vigna aconitifolia*), *Haramah* and *Paharimah* (Himachal Pradesh) or *Bejiamah* (Assam), *Nagamah* (Arunachal Pradesh), *Bete* (Mizoram), *Jami* and *Agukzungken* (Nagaland) and *Chak hawai* (Manipur) (Gautam *et al.* 2007). In addition, there are a number of local names for local varieties, but there is no authoritative account of what this boils down to in actual genetic diversity. The FOSRIN project has started a mapping of genetic diversity of more than 100 accessions from India and Nepal using molecular marker techniques, but the present evidence is far from conclusive even for these two countries.

The wild form *V. umbellata* var. *gracilis* has indeterminate growth and a twining habit and climbs tall on bamboos and trees. The seeds are found in many colours and sizes, from black or reddish-brown through grey-mottled to yellowish. Ricebean is quite closely related to Adzuki bean (*Vigna angularis*), which is traded internationally and has a large market in Japan.

The green pods are eaten as a vegetable although not very commonly. The dried seeds are used in various traditional dishes, but mainly cooked as a *dal*, a thick bean soup or sauce that is poured over rice or eaten with wheat *chapattis* and vegetable curries. It is consumed interchangeably with other pulses and can serve as a substitute for lentils, black gram, horsegram or pigeon pea. In many places it has a reputation as a poor man's food. The role of pulses in the diet is to a large extent as a partial substitute for animal source foods for the large share of the population in South Asia who on a day-to-day basis are predominantly vegetarian, whether for economic or for religious reasons. In my own judgement, ricebean is good food and more pleasant than several other pulses I have tasted. It can be used for *dal* soup/sauces and 'falaffel'-like items, as well as Brazilian *feijoada*.

The crop residue from ricebean is a valuable and palatable fodder which is known to increase milk production. It has been promoted in India as a cattle fodder (Mukherjee *et al.* 1980), and has also been suggested as a crop with potential for chicken fodder production. This might lead to an increased market for ricebean, but could also lead to the crop being stigmatized as 'livestock fodder'. Ricebean is easily established on exhausted soils. It is good at fixing nitrogen and is well suited for intercropping with local varieties of maize; in fact, it is producing much better intercropped with maize than as a monocrop (Rerkasem and Rerkasem 1988). Shorter varieties are grown on terrace risers on sloping land, providing food, fodder and protection against erosion. In Eastern Nepal it is at times used as green manure for mulching. Ricebean is relatively resistant to drought, pests and diseases. During a field visit in September 2009 close to Dharamsala in Himachal Pradesh, the FOSRIN team observed how many other crops were severely affected by a major, regional drought lasting about seven weeks, in addition to severe damages inflicted by wild boar and monkeys. Ricebean had remained dormant during the drought but flourished after the rains returned in the late monsoon, and was not affected by animals. During storage, ricebean may occasionally be attacked by bruchids, but less than several other pulses. Kashiwaba *et al.* (2003) tested 17 accessions of cultivated and wild accessions of ricebean and found that only 3 were not resistant to bruchids.

The nutritive value of ricebean is rather similar to other legumes. While the content of protein is not exceptional, its bioavailability is high. It has an excellent amino acid composition, is relatively rich in most B vitamins and is a good source of several minerals, particularly Ca, Fe, Zn, Cu and Mn (Andersen 2009). It contains some trypsin inhibitor[2] and polyphenols,[3] but apart from phytic acid in general, only modest amounts of the anti-nutrients that are a disadvantage of many legumes. Phytic acid is naturally reduced through overnight soaking. It has no allergenic or toxic compounds, and it contains less flatulence-producing saccharides than many other beans (Revilleza *et al.* 1990, Kaur and Kawatra 2000). In short, viewed as a crop and a source of nutrients, it has a number of advantageous characteristics. Chandel *et al.* (1978, p. 21) noted that 'the nutritive value of the rice bean is exceptionally high'.

The potential of ricebean to become a successful crop has been scientifically recognized for several decades (Chandel *et al.* 1978, Duke 1981), and historically it has spread from the original centres of cultivation in the Himalayas, China and Malaysia, further into China, Korea, Japan, Java, Mauritius, the Philippines, USA, Brazil and the West Indies (Duke 1981). However, it has not become an important crop anywhere, and the known international trade is very limited. No reliable data on changes in production and trade are available. The traditional farming practices – intercropping and planting on terrace risers – make it very difficult to map the acreage under cultivation, since the proportion of riceban compared to other intercrops may vary from field to field, and the conversion from vertical terrace risers to a horizontal map base is complicated. A survey carried out in four field sites in India and Nepal under FOSRIN (Andersen and Chandyo 2010) showed that ricebean was the sixth most commonly consumed grain pulse. These sites are not necessarily representative for the whole of the Himalayan hills, and they may have been impacted by extension activities under FOSRIN, so it is more likely that ricebean is less frequent in general. The anecdotal evidence appears to be one of decline everywhere. In Kathmandu, ricebean is normally available at the *Kalimati* market which serves as wholesale market for the Nepalese capital.

In Ilam district in Eastern Nepal, ricebean is still grown by many farmers. Ilam is reputed inside Nepal for a particularly successful development of agriculture. The basis of the production is a mixture of commercial dairy farming and horticulture in which the multiple ecological services of ricebean are realized: for food, cattle fodder, fixation of nitrogen, erosion protection of terrace risers, and not least as an intercrop with maize which is rotated with a main cash crop: ginger. Ricebean has a niche in Ilam since ginger cannot be grown on the same land 2 years in a row. Even so, ricebean is losing ground in competition with other legumes such as black gram and pigeon pea that are perceived to be tastier and fetching higher market prices. On the terrace risers, it must also compete with broomgrass (*Thysanchaena maxima*), which has an alternative dual purpose: broom production and fodder for dairy cattle, and has become one of the most important cash crops in Ilam. Although limited, there is a local market for ricebean, and it is being exported to the urban market in Siliguri in neighbouring India. There are no figures for this trade.

The farmers in Ilam listed a number of negative traits in the existing varieties of ricebean. The indeterminate growth habit[4] of most varieties is conflicting with the planting of subsequent crops. The time to maturity is commonly around 130–140 days. Unfortunately, most of the tastier varieties are those that mature late (noted by Khanal *et al.* 2007). Other problems are related to variability in seed size, seed hardness[5] and the twining habit of the plants, which tends to weigh down maize plants and makes it difficult to harvest. While the asynchronous flowering habit is an advantage for the use of green pods as a steady vegetable supply, it becomes a problem if the crop is grown to maturity, since a number of pods invariably will be full of small and immature seeds. Finally, the taste of ricebean was not the most preferred among the Ilam farmers. They said that they would be interested in improved varieties with determinate growth, large seeds and good taste qualities. Large seeds were generally preferred, but not always. One bold-seeded variety was seen as inferior in taste. The farmers in Ilam are well educated and have a fairly sophisticated understanding of soil nutrient management and all the ecological services of ricebean, but clearly expressed reluctance to continue growing ricebean if other crops are seen as more suitable.

Acharya (2008) did an in-depth study in Dang district in western Nepal. In contrast to Ilam, Dang is predominantly a poor area and the households studied are more subsistence oriented in their farming practices. In the past, ricebean was grown on a rather large scale, especially in intercropping with maize. Currently, the cultivation is declining. Interestingly, there are different reasons for dis-adoption between farmers. Among poor farmers, ricebean is being replaced by winter cultivation of staple crops. More wealthy farmers replaced it with cash crops, either vegetables or improved varieties of other pulses, especially black gram. In Dang, a preference ranking analysis placed the taste of ricebean as number four out of six pulses, after chickpea, black gram and lentil, but ahead of broad bean and pea. Among traders, it was put at the bottom of all pulses, since there is no export out of this region.

In Gulmi district, in the Middle Hills of Western Nepal, the cultivation and productivity of ricebean are decreasing due to increased cropping intensity, more emphasis on main crops, availability of newer pulse varieties and increased use of chemical fertilizer which lowers the productivity of ricebean (Khanal *et al.* 2007). Here, as in other places, there was dissatisfaction with hard seededness of some varieties. One landrace, *Bhadaure*, was described as 'not tasty because it is thin and has a strange taste' (Khanal *et al.* 2007) However, while the concept *taste* contains elements that may be of a general nature, the liking of a food may also be culturally constituted, and Neog *et al.* (2007) in North East India found that ricebean was the most preferred pulse food in Nagaland for taste as well as nutrition, while not so popular in Manipur.

In one location in Himachal Pradesh in India (own field observation in 2007), there was a major decline in cultivation of ricebean. An important constraint here was the build-up of blister beetle populations. After 2–3 years of ricebean cultivation, there was hardly any yield since the flowers were eaten by several species of blister beetles. The blister beetle problem was also observed in Assam in Eastern India. Blister beetles can be collected by hand, using plastic bags or gloves for protection, or they can be controlled with insecticides. However, these actions require additional labour or spending, and my observations from sites in India and Nepal indicate that farmers were not prioritizing the ricebean crops even when large numbers of blister beetles could have been killed easily. From Dang in Nepal, Acharya (2008) noted that farmers were not investing time or money in ricebean due to its relatively low economic return.

Attempting some generalisations from the case of ricebean

While ricebean and other underutilized crops are subject to the influence of factors belonging to the value chain, and competition with improved seeds of the major crops, the use or dis-adoption of the crop is ultimately the decision of the farmers. The experiences from different sites show that this may be linked to different reasons between subsistence and more commercially oriented farmers. By comparing the cases from Dang and Ilam, it is seen that poverty and pressure on the land may lead to dis-adoption as among poor farmers in Dang, while the under-utilized crops may find roles in highly commercial agriculture as in Ilam. This counters the hypothesis that modernisation inevitably leads to genetic erosion and extinction of minor crops, in agreement with the global assessment from FAO (2010).

However, the market price relative to alternative crops will always be a decisive factor for the farmer. Therefore, the success of ricebean will eventually depend on developing and making available seeds of varieties that are less variable, have determinate growth, good yield, and which have good organoleptic qualities and cooking properties. It is inevitable that this will lead to the reduction of some of the existing landraces, since these are inferior in one or more of these aspects. The experiences with IPR-related rules restricting flow of germplasm and the EU Novel Food regulations show that these types of regulations may have effects that can be contrary to their intentions. Crop development and trade is essential to the future of the underutilized crops.

If successfully turned into a commodity of higher value, ricebean may add to economic sustainability and food security of farmers. These are basically the aims of the FOSRIN project.

When these objectives have been fulfilled, ricebean has a potential to be one more pillar to lean on in terms of agricultural sustainability and food security. However, it is not a silver bullet which can dramatically improve nutrition and livelihoods among all marginal farmers. The abandonment of one traditional pulse crop does not necessarily lead to reduced agro-biodiversity at the species level, if it is being replaced by for instance black gram or soybean. Neither does it necessarily lead to impoverished nutrition. Maikhuri *et al.* (1997, cited from: Nautiyal *et al.* 2008) found in a study from central Himalaya that *Vigna* spp. (that may include ricebean) were being replaced by pigeon pea and amaranth. This can hardly be seen as a threat to local nutrition. The role of ricebean will often be as one potential substitute for other, rather similar choices of pulse crops. The overall problem of 'hidden hunger' and reduction of nutrient density (Bouis *et al.* 2000) is a serious issue, but should be viewed in a complex food system perspective. However, food diversity as such does not necessarily guarantee a completely balanced diet.

I will argue that some of the lessons from the work on ricebean can be generalised to other under-utilized crops. The statement by Padulosi and Hoeschle-Zeledon (2004, p. 6), that 'most underutilized species require some degree of improvement' may be an understatement. Rather, considerable effort is required in order to make the species in question competitive with major crops. While I agree that 'rational efforts that draw collectively on dietary diversity and biodiversity, rather than being dismissed as romantic, should be recognized as immediately pragmatic and ultimately essential' (Johns and Sthapit 2004, p. 152), the experience with ricebean shows the importance of acknowledging the views and aspirations of the local farmers and consumers. Under-utilized species are normally under-utilized for one or more reasons. These may be linked to farming system problems, such as being cumbersome to harvest or being prone to bird attacks and therefore requiring more manpower while maturing. But they may also be linked to the fact that the farmers are growing plants that not only are *crops*, but also must be understood as *food*. So, issues linked to cooking and consumption become paramount, such as ease of preparation, pleasant appearance and organoleptic qualities, while factors causing indigestion or flatulence, toxic and allergenic substances and unpleasant taste from for instance tannins and saponins may lead to abandonment, an issue also addressed by Smil (1997). Many of the existing varieties of under-utilized crops suffer from negative traits in this respect. The lessons from ricebean also suggest that there are dilemmas that must be acknowledged with respect to *in situ* conservation of within-species diversity. If some varieties of ricebean do not contribute to farmers' economic sustainability, they will understandably 'vote with the hoe' and do away with undesired biodiversity. Thereby, conservation of agro-biodiversity cannot be a responsibility of poor farmers if they can find better alternatives, o the general task of genetic conservation should be left to national and global organizations through off-farm strategies.

Acknowledgements

The FOSRIN project was funded by the European Commission under the 6th Framework Programme contract 032055. However, the opinions expressed are those of the author and not necessarily those of the FOSRIN consortium. I am indebted to Associate Professor Prem S. Chapagain, Central Department of Geography, Tribhuvan University, Nepal, for assisting with the fieldwork in Ilam, and the farmers who kindly shared their opinions. Finally, I have received encouraging and constructive comments on the manuscript from three anonymous referees and the editors of this volume of IJAS.

Notes

1. See http://www.ricebean.org
2. Trypsin inhibitors are plant compounds that counteract the human enzyme, trypsin, and thereby inhibit the metabolism of proteins. This inflicts on digestion.

3. Polyphenols may act as anti-nutrients by reducing iron availability in foods; on the other hand, they may reduce the risk of cardiovascular diseases and cancer.
4. Indeterminate varieties keep on growing and get new flowers until an external factor terminates growth.
5. Seed hardness refers to seeds that stay hard after soaking and therefore require longer cooking time.

References

Acharya, B.K., 2008. *Cultivation and use of ricebean. A case study of Dang District, Nepal* [online]. Thesis (M Phil), Norway: University of Bergen. Available from: https://bora.uib.no/bitstream/1956/2768/1/Masterthesis_ Acharya.pdf [Accessed 26 March 2012].

Andersen, P., 2009. Nutritional qualities of ricebean. *In*: P.A. Hollington, ed. *Food Security through Ricebean Research in India and Nepal (FOSRIN) Report 3* [online]. Bergen, Norway: Department of Geography, Universitet Bergen and Bangor, Wales, UK, CAZS Natural Resources, College of Natural Sciences, Bangor University. Available from: www.ricebean.org [Accessed 26 March 2012].

Andersen, P. and Chandyo, R.K., 2010. Health and nutrition impacts of ricebean. *In*: P.A. Hollington, ed. *Food Security through Ricebean Research in India and Nepal (FOSRIN) Report 6* [online]. Bergen, Norway: Department of Geography, Universitet Bergen, and Bangor, Wales, UK, CAZS Natural Resources, College of Natural Sciences, Bangor University. Available from www.ricebean.org [Accessed 26 March 2012].

Bouis, H.E., Graham, R.D., and Welch, R.M., 2000. The Consultative Group on International Agricultural Research (CGIAR) micronutrients project: justification and objectives. *Food and nutrition bulletin*, 21 (4), 374–381.

Chandel, K.P.S., *et al.*, 1978. Rice bean – a pulse with high potential. *Indian farming,*, 28, 19–22

Chishakwe, N.E., 2008. *An overview of the international regulatory frameworks that influence the conservation and use of underutilized plant species.* Rome, Italy: Global Facilitation Unit for Underutilized Species, and Nairobi, Kenya: the Genetic Resources Policy Initiative.

Duke, J.A., 1981. *Handbook of legumes of world economic importance.* New York and London: Plenum Press.

FAO, 2010. *The second report on the state of the world's plant genetic resources for food and agriculture.* Rome: Food and Agriculture Organization of the United Nations (FAO, Commission on genetic resources for food and agriculture).

Fassil, H., *et al.*, 2000. Diversity for food security: improving human nutrition through better evaluation, management, and use of plant genetic resources. *Food and nutrition bulletin*, 21 (4), 497–502.

Gautam, R., *et al.*, 2007. *Distribution of ricebean in India and Nepal* [online]. Pokhara, Nepal: Local Initiatives for Biodiversity, Research and Development and Bangor, Wales, UK, CAZS Natural Resources, College of Natural Sciences, Bangor University, Food Security through Ricebean Research in India and Nepal (FOSRIN) report 1. Available from: www.ricebean.org [Accessed 26 March 2012].

Gruère, G., Giulani, A., and Smale, M., 2006. *Marketing underutilized plant species for the poor: a conceptual framework.* Washington: IFPRI, EPT Discussion Paper 154.

Hermann, M., 2009. The impact of the European Novel Food Regulation on trade and food innovation based on traditional plant foods from developing countries. *Food policy*, 34, 499–507.

Johns, T., 2004. Underutilized species and new challenges in global health. *LEISA Magazine*, March issue, 7.

Johns, T. and Sthapit, B.R., 2004. Biocultural diversity in the sustainability of developing-country food systems. *Food and nutrition bulletin*, 25 (2), 143–155.

Kashiwaba, K., *et al.*, 2003. Characterization of resistance to three bruchid species (*Callosobruchus* spp., Coleoptera, Bruchidae) in cultivated rice bean (*Vigna umbellata*). *Journal of economic entomology*, 96 (1), 207–213.

Kaur, M. and Kawatra, B.L., 2000. Effect of domestic processing on flatus producing factors in ricebean (*Vigna umbellata*). *Nahrung*, 44 (6), 447–450.

Khanal, A.R., *et al.*, 2007. *Farmers' local knowledge associated with the production, utilization and diversity of ricebean (Vigna umbellata) in Nepal* [online]. Pokhara, Nepal: Local Initiatives for Biodiversity, Research and Development, Rampur, Chitwan, Nepal: Institute of Agriculture and Animal Sciences, and Bangor, Wales, UK: CAZS Natural Resources, College of Natural Sciences, Bangor University, Food Security through Ricebean Research in India and Nepal (FOSRIN) report 3. Available from: www.ricebean.org [Accessed 26 March 2012].

Maikhuri, R.K., *et al.*, 1997. Eroding traditional crop diversity imperils the sustainability of agricultural systems in Central Himalaya. *Current science*, 73, 777–782.

Mukherjee, A.K., Roquib, M.A., and Chatterjee, B.N., 1980. Rice-bean 'K-1' for the scarcity period. *Indian farming*, 29, 19–20.

NBPGR, 2007. *State of plant genetic resources for food and agriculture in India (1996–2006). A country report*. New Delhi: National Bureau of Plant Genetic Resources, Indian Council of Agricultural Research.

Neog, S.B., Bora, R., and Mili, R., 2007. *Preferred Trait Analysis (PTA) for ricebean in North Eastern Hill states in India, 2007.* Unpublished FOSRIN Technical Report.

Padulosi, S., Eyzaquirre, P., and Hodgkin, T., 1999. Challenges and strategies in promoting conservation and use of neglected and underutilized crop species. *In*: J. Janick, ed. *Perspectives on new crops and new uses*. Alexandria, VA: ASHS Press, 140–145.

Padulosi, S. and Hoeschle-Zeledon, I., 2004. Underutilized plant species: what are they? *LEISA Magazine*, March issue, 5–6.

Ravi, S.B., *et al.*, eds., 2006. *Hunger and poverty: the role of biodiversity – report of an International Consultation on The Role of Biodiversity in Achieving the UN Millennium Development Goal of Freedom from Hunger and Poverty.* Rome, Italy: M.S. Swaminathan Research Foundation, India/ IPGRI and GFU for Underutilized Species.

Rerkasem, K. and Rerkasem, B., 1988. Yields and nitrogen nutrition of intercropped maize and ricebean *(Vigna umbellata* [Thumb.] Ohwi and Ohashi). *Plant and soil*, 108, 151–162.

Revilleza, M.A.J.R., Mendoza, E.M.T., and Raymundo, L.C., 1990. Oligosaccharides in several Phillipine indigenous food legumes: determination, localization and removal. *Plant foods for human nutrition*, 40, 83–93.

Shiva, V., 1993. *Monocultures of the mind. Perspectives on biodiversity and biotechnology.* London/ New York/Malaysia: Zed Books.

Shiva, V., 2000. *Stolen harvest. The hijacking of the global food supply.* Massachusetts: South End Press.

Smil, V., 1997. Some unorthodox perspectives on agricultural biodiversity. The case of legume cultivation. *Agriculture, ecosystems and environment*, 62, 135–144.

Toledo, Á. and Burlingame, B., 2006. Biodiversity and nutrition: a common path toward global food security and sustainable development. *Journal of food composition and analysis*, 19, 477–483.

Crop–livestock systems in rural development: linking India's Green and White Revolutions

Pratyusha Basu[a] and Bruce A. Scholten[b]

[a]*Department of Geography, Environment and Planning, University of South Florida, 4202 E. Fowler Ave., NES 107, Tampa, FL, 33620, USA;* [b]*Department of Geography, Science Laboratories, Durham University, South Road, Durham, DH1 3LE, UK*

In discussions on agricultural transformations produced by Green Revolution technologies, an exclusive focus on crops has led to incomplete understanding of the value of livestock in rural economies. The incorporation of crop–livestock systems seems even more pertinent in the context of India, given that the Green Revolution here was also accompanied by a White Revolution in dairying. This paper explicates the links between India's Green and White Revolutions by juxtaposing international and national policies of rural development with village-level practices of cropping and dairying. It begins by comparing the Green and White Revolutions in India, both the dissimilarities that marked their origins and how the White Revolution subsequently grappled with a notion of productivity mirroring that of the Green Revolution. Two village-level case studies then illustrate how dependence on both crops and cattle shape rural livelihoods in ways that require an interrelated understanding of Green–White Revolutions. Overall, this paper shows how the socio-economic and environmental value of crop–livestock farming is valuable for understanding the complexities of small-scale livelihoods, and suggests lessons for building sustainable programmes of rural development.

Introduction

In rural development programmes, a tendency to equate agriculture with crops has often led to an exclusive focus on plant species and their role in rural landscapes and livelihoods, and this is exemplified by the Green Revolution based on high-yielding variety (HYV) seeds. In the process, the significant role of animals in agriculture is often ignored (Devendra and Thomas 2002, Moritz 2010). Yet, across rural spaces, crops and animals are often interrelated in practices of agriculture (Devendra 2002, Kurosaki 2006, Herrero and Thornton 2010). Such interrelations become especially noteworthy in the context of India where the Green Revolution occurred concurrently with the White Revolution in dairying.

The aim of this paper is to provide an interconnected perspective on India's Green and White Revolutions, terms used to denote the marked rise in foodgrain and milk productivity, respectively, over the 1960s–1980s. It provides a framework for understanding how sustainable rural livelihoods are constructed in local contexts through a combination of crops and livestock.

Such interlinked analysis of the Green–White Revolutions is important at this juncture since India is increasingly being promoted as a model to be emulated in terms of rural development (Stringer *et al.* 2008, Toenniessen *et al.* 2008, World Bank 2012a, 2012b). In the process, this paper seeks to make two contributions to contemporary analysis of agricultural sustainability and rural development. First, it examines the productivity–participation dichotomy as it appears within Green and White Revolution policies, illustrating the ways in which productivity has risen to the status of paramount goal of rural development (Yapa 1993, Gupta 1998). The adoption of improved technologies in order to meet the goal of higher production has been the central characteristic of the global Green Revolution. In terms of India's White Revolution, the main objective of national dairy development was the organization of milk collection, processing and especially marketing around participatory cooperative structures (Kurien 1997, Scholten 2010). However, the programme also privileged higher milk yields through improved cattle, and ultimately this imperative may have contradicted its emphasis on local participation (Rajapurohit 1979, Basu 2009a). The values attached to participation and productivity thus become an important issue for understanding the intersections of the Green and White Revolutions.

Second, this paper seeks to identify the ways in which the success of the Green Revolution cannot be considered in isolation from the inputs provided by the White Revolution and vice versa. Thus, small farmers in India ensure the sustainability of both the Green and White Revolutions through crop–livestock interactions, despite separate programmes directed at crop and milk yields. Further, crop–livestock interactions demonstrate an ethos of self-sufficiency, as locally available human and environmental resources combine to mitigate excessive dependence on market-based inputs. A focus on village-level practices thus balances an exaggerated notion of the power of official policies, and ensures that lessons can be learned from the experiences of small farmers in India (George 1985, Basu 2009b). It should be noted, however, that this paper is not arguing that crop–livestock systems are the only way in which sustainable farming can be undertaken. Instead, the argument is that already existing crop–livestock systems should be valued and protected instead of being discarded in the latest round of agricultural development.

The next section situates the concerns of this paper within the broader context of political economy, critical development and sustainable approaches to agriculture. After that, the paper traces the institutional histories of Green and White Revolutions in India, showing how they meshed despite potential differences in ideological underpinnings. The links between cropping and dairying in two case study villages are examined next in order to understand locally divergent experiences of the Green and White Revolutions. The paper concludes by considering the implications of experiences within India for replicating Green and White Revolutions on other continents and thus seeks to extend contemporary debates over the possible futures of rural development.

Small farmers and sustainability

> Smallholder farming has been long neglected. It is not a single solution, but an important component of both hunger and poverty reduction (Foresight 2011, p. 25).

This paper's focus on mixed farming seeks to contribute to three main issues in contemporary rural development: explaining persistence of small-scale farming, enabling participatory development and constructing feasible strategies for sustainable intensification of agricultural production. The importance of small-scale farming for the sustainability of rural communities is being increasingly recognized by international development agencies and within agrarian and rural development studies (Jayne *et al.* 2010, Wiggins *et al.* 2010, Viswanathan *et al.* 2012). This is partly in response to the puzzle of the persistence of smallholders across much of rural Asia and

Africa, despite mainstream pronouncements on the inevitability of their incorporation into urban economies due to the increasing commercialization of agriculture (World Bank 2007, Hazell 2011). This paper argues that mixed farming provides one way in which to explain this puzzle of persistence. For instance, in the context of India[1] – even as Green Revolution technologies tended to be more easily adopted by large farmers – it is likely that the simultaneous implementation of Green and White Revolutions ensured that synergies between crops and livestock could be used to strengthen the overall household economies of small farmers. While this persistence of small-scale holding has often been viewed as an obstacle to the full unleashing of the efficiencies of Green Revolution technologies, it also points to the possibility of building spaces of subsistence in the interstices of large-scale cultivation, spaces that are likely to have been crucial to providing a material base for the eventual diversification from farm to non-farm incomes. Mixed farming on small farms therefore is one important way in which to extend studies of the class-differentiated outcomes of the Green Revolution.

Contemporary development is often structured around a participatory framework, and practices of mixed farming provide a way in which to critically rethink the meaning of participation (Cornwall 2002, Hickey and Mohan 2004). Thus, existing approaches seek to enable participation by providing spaces within which development officials can meet with community members targeted for development interventions. This assumption, that spaces of dialogue are sufficient to participation in development, ignores how rural development programmes are incorporated into existing rural practices by farmers themselves, so that participation is a process that occurs outside the gaze and control of development officials, rather than only in terms of interactions between officials and community members. As the case study for this paper shows, crop–livestock farming is practised in rural India not due to any overt attempt on the part of rural development programmes to combine the two, but mainly due to the ability of farmers to combine development inputs across different programmes. The notion of 'mixed' therefore reflects a particular form of participation, one that emerges not at the expert–community nexus, but in the ways in which a set of diverse, if not divergent, development policies are fitted into local places.

The competition between agricultural and environmental functions of land has also led to a call for new strategies whereby food production can be pursued without simultaneously losing ecosystem values, or intensifying environmentally and socially harmful use of chemicals (The Royal Society 2009). The new call is for 'sustainable intensification', seeking an amalgam of economic, social and environmental sustainability of agriculture. Utilization of crop–livestock interactions, or fully exploiting the use of agricultural residues as fodder and animal dung as fertilizer, could be one way in which crop–livestock systems contribute to economic and environmental efficiencies and hence to the goal of sustainable production. Mixed farming could thus be implemented as a logical outcome of the notion that sustainability is predicated on diversity. As the case study villages of this paper show, the combination of Green and White Revolutions ultimately strengthened both crop and dairy farming in the two villages, so that sustainable rural livelihoods were produced at the intersection of small farms and local practices of cropping and dairying, rather than due to any specific attempt to combine Green and White Revolutions at the policy level.

Productivity and participation: linking Green and White Revolutions

The Green and White Revolutions in India emerged from very different approaches to development. This section outlines how such differently imagined programmes ultimately began to resemble one another in terms of their goals and the problems they faced. The HYV seeds which propelled the rise in productivity associated with the Green Revolution were devised primarily by US agricultural scientists through a combination of controlled approaches to plant breeding, and public and private funding for international development (Parayil 1992). Global dissemination of HYV seeds occurred under the aegis of American development agencies, subsequent to the popularization of development as technology transfer from more to less developed economies (Escobar 1995, Cowen and Shenton 1996). International development could thus display American benevolence, imperative in a context where capitalist development was involved in a Cold War with socialist development, the latter often being the preferred ideology for newly independent countries (Finnemore 1997). Green Revolution agricultural techniques were introduced in India in the late 1960s, and HYVs were especially extolled as basic to food self-sufficiency at the national level and key to avoiding a Malthusian crisis, even as critical perspectives pointed to the class inequalities that characterized this programme (Gupta 1998).

The trajectory of India's White Revolution was diametrically opposed to that of the Green Revolution, and the cooperative dairying model on which the White Revolution was based predated the Green Revolution by a couple decades. Thus, cooperative dairying was established on the eve of India's independence in the aftermath of a local struggle against private control of milk marketing (Kurien 1997). In 1946, after British colonial authorities granted the Polson company monopoly rights to milk procurement in Kaira district (currently Kheda and Anand districts in the state of Gujarat in western India), farmers around the small town of Anand organized a boycott of Polson. They drew strength by linking their struggle to the national independence movement. The success of this boycott forced the British government to rescind Polson's monopoly, and the farmers of Anand went on to form the Kheda District Cooperative Milk Producers Union (KDCMPU), popularly known by its brand name, Amul. Links with nationalist leaders proved valuable to KDCMPU, facilitating access to milk contracts and prominence on India's rural development landscape.

The success of KDCMPU could also be related to its establishment in a region that was already prosperous. The long history of agricultural development in this region of central Gujarat was linked to British colonial policies that favoured settled agricultural practices and social groups linked to them (Rutten 1995). Cooperative organization was also supported by the cohesiveness of the Patidar caste, whose members comprise the majority in the present-day districts of Kheda and Anand, and whose caste solidarity had been fostered through an adherence to marital exchanges organized around customarily grouped villages (Basu 2009b). Cooperative dairying was thus part of a farmer-led strategy to enhance agricultural incomes by diversifying into dairying, so that crop–livestock farming can be considered a pre-Green Revolution strategy for economic growth in central Gujarat.

In the early 1950s, KDCMPU obtained a contract to supply milk to the city of Bombay (now Mumbai), and this access to a large urban market, coupled with the fact that Anand was well connected by rail transport networks, ensured the steady growth of cooperative dairying. By 1960, however, the erstwhile state of Bombay was divided into the states of Maharashtra and Gujarat, and KDCMPU's location in Gujarat jeopardized access to the markets of Bombay city which had gone to the state of Maharashtra (Patel 1990). This changed configuration of state boundaries partly led KDCMPU to seek the attentions of India's central government. In 1964, then Prime Minister Lal Bahadur Shastri, known as sympathetic to rural causes, visited KDCMPU and spent time in a village near Anand. This visit impressed Shastri sufficiently, and led the central government to designate rural cooperative dairying as the sole model of dairy development in India and establish the National Dairy Development Board (NDDB) in Anand in 1965. Cooperative dairying was now folded into the embrace of national development.

The primary task before the NDDB was to replicate cooperative dairying across rural India, a programme designated as Operation Flood (Doornbos *et al.* 1990, Scholten 2010). This required funding and led the NDDB to international development agencies. Dairy aid from the European Economic Community was one of the first major sources of funding and the NDDB is credited with insisting on monetization of this dairy aid whereby revenue obtained from sale of European butter oil and milk powder in India's major metropolises was used to fund personnel and infrastructure for replication of cooperative dairying. Food aid was thus channelled towards developing, rather than competing with, India's dairy farmers (Scholten 2010). By the late 1960s, funding for the Green Revolution had also gathered strength, and soon the ideologies of the Green and White Revolutions began to intersect.

This intersection can be partly traced to the increasing prominence of the World Bank in cooperative dairying (Candler and Kumar 1998). In 1974, the World Bank provided a major loan for the replication of rural dairy cooperatives, pushing the profile of Operation Flood higher in the realm of international development. Simultaneous with financial support, the Bank promoted the high-yielding crossbred cow, a cross between European dairy breeds and the indigenous zebu breed. Candler and Kumar (1998) quote internal Bank documents from 1969, which show how support for crossbred cows emerged from a desire to approximate the productivity of the Green Revolution. In the words of Robert McNamara, then President of the World Bank,

> The future lies with the hybrid cow – in the beginning a first cross of European milk breed (mainly Jersey) on the native cow which in one generation will double production and about halve costs .... It's the most exciting cattle development prospect in the world at the present time. ... In my view, the prospect is the animal equivalent of the rice and wheat story with the possibility of even greater initial gains and impact upon costs of production which, with the buffalo, can only remain high (Candler and Kumar 1998, p. 3).

This informal memo reflects a tendency within the World Bank to extol the virtues of dairy cows over dairy buffaloes, even as the production figures mentioned in the memo were highly tentative. In this account, international funding seemed to lead cooperative dairying towards borrowing from the narrative of hybrid productivity. In the words of a critic of this strategy of cattle improvement, this led to a situation where '[w]hilst other tropical countries import Indian cattle to breed their animals stronger and hardier, India herself [would be] ... frantically importing European breeds to make her zebus more delicate and demanding' (George 1985, p. 124).

HYV seeds and crossbred cows thus became central figures for building a comparative understanding of the Green and White Revolutions and, in critical analyses of the Operation Flood programme, two problematic aspects of their relationship were highlighted. First, one of the most important uses of cattle in India is as draft animals for agricultural work, including ploughing, weeding and transport (Rosenzweig and Wolpin 1993). While mechanization is often cited as a feature of the Green Revolution, such draft uses of cattle have not completely disappeared in rural India. Phaniraja and Panchasara (2009, p. 404) estimate that over '55% of the total cultivated area is still being managed by using draught animals as against about 20% by tractors'. However, with the implementation of the second phase of the Operation Flood programme in 1981, there was increased emphasis on genetic improvement of dairy animals in India through crossbreeding. Crossbred cattle were not suitable draft animals and could only be used as dairy animals. The popularization of crossbred cattle thus could have deleterious consequences in terms of access to draft cattle for agricultural work (George 1990).

Second, shortage of fodder was noted as one the principal constraints to increasing dairy productivity in India (Singh 1979, Nair 1980). The higher yields of crossbred cows leads to a corresponding increase in demand for animal feed and increased productivity through the Green Revolution was key to satisfying this need (Nair 1985). Yet, to the extent that hybrid varieties were less palatable to dairy animals, or not traditionally fed to them, Green Revolution crops had produced the opposite effect in some contexts (George 1990). For instance, in the Malwa region of the state of Madhya Pradesh, which was one of the sites where cooperative dairying had been replicated on the basis of World Bank funding, farmers had shifted to cultivation of soybean, and since all the soybean was destined for export, there was no excess available for use as animal feed (Hiremath *et al.* 1997). Plus, farmers there did not prefer to feed soybean to their dairy animals. Changes in cropping patterns consequent to the Green Revolution thus were not always conducive to dairying, and the replication of cooperatives suffered serious setbacks where this mismatch occurred.

To increase fodder availability, one strategy promoted under the second phase of Operation Flood was to shift agricultural land to cultivation of green fodder. This, however, was problematic in part because

> [t]he energy balance of this specialised model in the food system [whereby land dedicated to fodder production is increased] will be negative since most of the resources used here can be used directly for producing food for human consumption. ... This model stands in contra-distinction with the 'integrated model' of dairy development that has evolved in India. From the point of view of maintaining the energy balance in the food system it is important to move away from the 'western model' to an effective 'integrated model' (Nair 1985, p. A-95).

Thus, Nair argued that competition over devoting land to food or fodder could be one of the limiting factors of both Green and White Revolutions. The need to focus on synergies between crop and livestock systems becomes especially clear within this perspective.

Similar to questions raised about limits to productivity, the nature of participation in Green and White Revolutions has also been much scrutinized, especially in terms of class and gender inequalities. Studies of the Green Revolution have pointed out that new technologies were more easily adopted by farmers with relatively large holdings, so that it accentuated rural class inequalities (Byres 1981, Dhanagre 1987, Das 2002, Herring 2007, Birkenholtz 2009). Cooperative dairying offered a model of rural development that, unlike the Green Revolution, would combine economic and social goals (Kurien 1997). Given that small farmers comprise a majority of the membership of cooperatives (e.g. Amul 2010), the economic value of dairying for small-scale livelihoods is clear. Yet, entrenched rural class divisions cannot be considered to have been dislodged, and leading dairy officials were quick to point out that cooperative dairying cannot be an 'all-purpose poverty alleviation programme' (Katakam 2002).

The implications of the Green and White Revolutions for women's work burdens have also been widely discussed. The Green Revolution is connected to an intensified use of women's labour, especially of women from small landholding and landless families (Agarwal 1984, Chowdhury 1992, Ramamurthy 1997, 2011, Srivastava and Srivastava 2010). Cooperative dairying, particularly where dairying is part of women's traditional responsibilities, is also linked to increases in women's work (George 1985). While the economic benefits of increased agricultural productivity have not been directly available to women, partly due to women's exclusion from land ownership, cooperative dairying has often provided women with direct access to a steady source of income which constituted the principal difference between the Green and White Revolutions. However, there were also similarities between the two and these were most apparent in the case of loss of women's knowledge. Shiva (1988) provides the most extensive critique of the truncation of women's knowledge of seeds and indigenous plant varieties due to the Green Revolution. Her critical analysis also draws on George's (1985) discussion on the extent to which cooperatives reduced women's roles in dairying. Thus, women's milk processing work has been taken over by dairy plants reducing local availability of milk by-products. Moreover, neither the Green nor White Revolutions have radically raised the social status or political clout of women. In the case of the Green Revolution, this becomes apparent in the social tendency to devalue women's agricultural work through labelling it as domestic work (Chowdhury 1992).

In the White Revolution, women's nominal presence in village-level dairy boards rarely translated into any actual control over the administration of cooperatives (George 1985). Some steps are being taken to remedy this situation, such as the establishment of women-only dairy cooperatives in contexts where cooperative dairying was replicated (Candler and Kumar 1998). But social inequalities will remain a matter of debate especially when success is mainly measured in terms of productivity.

The Green and White Revolutions emerged under very different conditions – the former an internationally coordinated scientific endeavour, the latter a locally based farmers' movement – but began to resemble each other due to the overwhelming hegemony of productivity in official rural development circles. Plus, similar to the Green Revolution, the White Revolution became widely considered a transferable, or replicable, model of development. In the process, notions of local control were subsumed under the idea that cooperatives, even those imposed by official decree, were inherently democratic. Currently, rural development in India is conducted under the global turn towards neoliberal economic policies (Oza 2006, Moore 2008), which devalue state support, favour privatization and prioritize productivity over participation. This adds another layer to the homogenization of the rural policy landscape. In the case of the Green Revolution, neoliberal policies threaten subsidies associated with seeds, fertilizers, pesticides, irrigation and electricity, even as they enable some farmers to gain access to world markets. Where the White Revolution is concerned, the role of rural cooperative dairying as the exclusive model of dairy development is under threat (Scholten and Basu 2009), partly signalled by the withdrawal of World Bank funding in 1996 under the rationale of 'levelling the playing field for agroindustrial development between cooperative and private entrepreneurs' (Candler and Kumar 1998, p. 12).

A focus on official policies of the Green and White Revolutions however obscures the village-level work associated with building links between seeds and cattle, cropping and dairying. Thus, the somewhat floating notion of development as embodied by national and international agencies needs to be grounded in material practices to understand how rural development is actually experienced. The next section focuses on how the link between the White and Green Revolutions, which emerges at the official level in terms of a shared notion of productivity, is shaped locally through symbiotic practices of crop farming and dairying geared more towards the sustainability of rural households.

Mixing crops and dairying: village perspectives on Green and White Revolutions[2]

This section focuses on two case study villages – Mahdol (pseudonym, located in the state of Gujarat) and Saundara[3] (pseudonym, in the state of Madhya Pradesh). The Green and White Revolutions have been more successful in Gujarat than in Madhya Pradesh and, during field research, the markedly better quality of roads in rural Gujarat was visual evidence of the state's greater economic prosperity. The two villages were chosen to reflect differences in cooperative dairying between the region where the programme originated (Gujarat) and sites where it was subsequently replicated (Madhya Pradesh, see Mergos and Slade 1987, Hiremath *et al.* 1997). In the process, these villages also became valuable case studies of the links between crop and dairy farming.

This section draws on data collected over 2000–2001 through household surveys, individual interviews and participant observation (Basu and Chakraborty 2008, Basu 2009b). The time period of data collection is significant because the effects of increased privatization of the agricultural sector due to the 1990s economic liberalization in India were yet to disturb village-level economies. This study does not aim at providing a comprehensive analysis of development programmes in the two villages, but aims to focus on the ways in which crop–livestock systems have been constructed through place-specific combinations of Green and White Revolutions. However,

Table 1. Comparing case study villages.

	Mahdol, Anand district, Gujarat	Saundara, West Nimar district, Madhya Pradesh
Population		
Number of households	277	469
Agriculture		
Total agricultural land	473 acres	1,407 acres
Community grazing land	None	15 acres
Households owning land	174 (63%)	164 (35%)
Total number of bullocks	16	268
Access to irrigation	1950s (irrigation cooperative); 1970s (Mahi canal)	1980s (private pipelines from Narmada River)
Number of tractors	12	13
Dairying		
Dairy cooperative established	1961	1986
Cooperative member households	190 (69%)	57 (12%)
Daily milk collected (approx.)	2,000 litres	400 liters
Total number of cows	370 (crossbred cows)	130 (Zebu)
Total number of female buffaloes	145	340

as described in some detail below (also Table 1), the two villages become useful for comparison since both have well-developed agricultural economies but do not show a similar level of participation in cooperative dairying. Further, while the villages are continuing to transform in accordance with new economic policies, this study seeks to focus on a specific moment when dairy cooperatives were functioning through crop–livestock interactions, in order to underline the value of mixed farming. The two case study villages broadly resembled one another in terms of experiences with the Green and White Revolutions. This is partly because both had access to irrigation: Mahdol due to canals drawing water from dams on the Mahi River, and Saundara due to its location along the Narmada River. However, unlike the state-led irrigation project in Mahdol, Saundara's irrigation was dependent on privately constructed pipelines. Both villages have successful dairy cooperatives, although Mahdol's encompassed a larger percentage of village households and showed much higher production (Table 1). Differences between the two villages are apparent when the details of crop and dairy farming are considered. Thus, the crossbred cow is the principal dairy animal in Mahdol, while Saundara is dependent on water buffaloes. While women in Mahdol are mainly responsible for dairy work, in Saundara, women mainly work field crops and often forego dairy activities. These contrasts form the basis of the discussion of crop–livestock farming in the rest of this section.[4]

Crop–livestock interactions can be measured in terms of the relationship between land and livestock ownership. Table 2 provides the results of bivariate parametric correlation of land ownership, dairy animal ownership and household size for the two villages using Pearson's correlation coefficients (r-values). In both villages, all three variables show a significant and positive association with each other. Land shows a stronger positive association with dairy animals in Saundara ($r = 0.735$) when compared with Mahdol ($r = 0.513$), which could indicate greater dependence on cooperative inputs in Mahdol. Household size can serve as a proxy for availability of labour. Correlations between dairy animals and household members in both Mahdol and Saundara are similar in value, thus masking differences in gender divisions of work between the two villages. The stronger positive relationship between land and household members in Saundara ($r = 0.495$) when compared with Mahdol ($r = 0.292$)

Table 2. Correlations between amount of land owned, number of dairy animals and number of household members.

	Mahdol		Saundara	
	Number of dairy animals	Number of household members	Number of dairy animals	Number of household members
Land owned (in acres)	0.513**	0.292**	0.735**	0.495**
Number of dairy animals	–	0.438**	–	0.423**

$^{**}p < 0.01$.

can potentially be attributed to the larger number of joint families in Saundara (Basu 2009b). These relationships can be further explained and clarified on the basis of ethnographic data.

Land and dairy animals

The Green Revolution in Mahdol and Saundara follows a long history of colonial and postcolonial agricultural development led by castes traditionally involved in settled agriculture. In both villages, multiple cropping is the norm and between two and four harvests are obtained annually. This intensification of agriculture can be traced to a combination of Green Revolution seeds and access to assured irrigation (Dubash 2002, Birkenholtz 2009). Mahdol's access to government-constructed irrigation in the 1970s was partly built on the region's political associations that had previously spurred the rise of cooperative dairying. Access to irrigation had also been organized much earlier through a village level irrigation cooperative formed in the 1950s. This cooperative ensures that a large part of small landowners in the village have access to irrigation at reasonable costs, and that private well owners do not inflate the prices at which they supply water. Cooperation is thus an ethos pervading the wider village economy, and is not just restricted to dairying. In contrast, access to irrigation in Saundara is based on about 40 privately owned pipelines (shared between 54 households, 12 percent of total households in the village), which draw water from the Narmada River. This reflects both the fact that 65 percent of the households in the village own no land, and that groundwater sufficient for irrigation is available given the proximity to the river. Pipelines therefore are an investment that become feasible only for large farmers often through kin-based partnerships.

The main crop in Mahdol is rice, and its cultivation began after the introduction of HYV seeds. Central Gujarat, the region in which Mahdol is located, is famous for its red soils which produce good-quality tobacco, but the poor quality of Mahdol's soils and the capture of the tobacco market by larger villages have reduced tobacco cultivation. Fodder crops are grown on a regular basis, including jowar (sorghum). Saundara is part of the black soils region of India which is especially suited to cotton. Other main crops include wheat and, since the advent of irrigation pipelines, sugarcane. Jowar is also cultivated here and used as both food and fodder. Access to seeds, fertilizers and pesticides in Mahdol and Saundara is facilitated by state-supported agricultural services cooperatives. The relative importance of agriculture in Saundara is reflected by the fact that elections to agricultural services cooperatives are hard fought, while dairy cooperative board members are chosen by consensus. In contrast, positions in the dairy cooperative are highly prestigious and more seriously contested in Mahdol.

As shown in Table 1, 63 percent of households own land in Mahdol as opposed to 35 percent in Saundara. This is partly due to the presence of river-dependent landless castes in Saundara. Cooperative dairy membership shows even greater difference, with 69 percent of households having membership in Mahdol and only 12 percent in Saundara. The link between agricultural land and dairying becomes more complicated when land ownership and dairy membership in the two villages are analysed in further detail. In Mahdol, the largest portion of cooperative members (46.3 percent) own only 1–2 acres of land. In Saundara, however, 45.6 percent of cooperative members own 6–10 acres of agricultural land. Small farmers thus represent a larger percentage of the dairy cooperative in Mahdol, and this can partly be explained by the stronger presence of the cooperative there. Small farm memberships also show a relative specialization in dairying in Mahdol, while dairying is linked to agriculture in Saundara.

In Mahdol, most landowning households own very small parcels of land for a number of reasons. Land has been sub-divided for generations, and the possibility of adding farmland is limited because the village was established in the interstices of nearby villages and is on the border of the town of Petlad. The pressure on land is apparent in the fact that there is no common grazing land in Mahdol. Small land parcels promote dairying since they are unprofitable for input-intensive Green Revolution crops but ideal for fodder. The cooperative dairying programme thus enables households to draw income from and retain ownership of small parcels of land. Farmers can also compensate for lack of access to land-based resources by buying cattle feed from the cooperative, paid for in part by their cooperative income.

The link between Green and White Revolutions, between crops and dairying, was mentioned in a number of interviews conducted in the village. As one farmer expressed it:

> In this village, no one owns more than 5 acres [94% of Mahdol's households own less than 5 acres, this farmer however owned more than 5 acres]. . . . That is why the Green Revolution is accompanied by dairying. These two complete one another, they are complementary [his word], because if you pursue only farming you cannot live on that, you cannot survive on that, which is why dairying proceeds in a complementary manner . . . (Interview, Mahdol village, March 2001).

The role of irrigation facilities, available as part of the spread of the Green Revolution, and dairying is also revealed in the connection made between the Mahi canal and dairy development by a prominent member of the dairy cooperative board.

> Pratyusha: Was there more change [economic development] because of the canal or because of the dairy cooperative?
>
> Dairy board member: If the canal was not there, we wouldn't be able to keep animals. Both were equal [in their contribution to economic development] (Interview, Mahdol village, March 2001).

In Saundara, there is a stronger association between land and buffalo ownership (Table 2). Dairying here is a subsidiary occupation, pursued to the extent that agricultural by-products (e.g. weeds, post-harvest residues) sustain cattle. Evidence that dairying income is subsidiary in Saundara is provided by the fact that members continued to sell milk to the cooperative even in a long period when the Indore Milk Union was not making regular payments. A loss could thus be borne by households in anticipation of future payments.

Dairy cooperatives in both villages have lower participation by landless households. In Mahdol, landless households comprised 23.2 percent of cooperative members, which supports the programme's claim that landless people benefit from cooperative dairying (e.g. Amul 2010). A landless labourer who works as bhagiya (cultivating land for a farmer in return for a share or bhag of the harvest), mentioned that,

> Half my income is from bhag, half is from dairy. Sometimes income from dairy is even more than from bhag. Patels [landed farmers] also earn more due to dairy, some people have become rich due to dairy (Interview, Mahdol village, June 2001).

However, landless membership was possible only through links with landed farmers (who employ them as labourers), so that cooperative participation was effectively controlled by landowners. The lack of common grazing land makes it difficult for the landless to source fodder. Moreover, high-yield crossbred cows are expensive to maintain for agricultural labourers, so income from dairying cannot become a significant source of their livelihood. In Saundara, only 3.5 percent of cooperative members were landless. A large part of landless groups in the village are fishers or boat owners who draw livelihoods from the river and not land. Their principal livestock ownership was goats which are grazed on the banks of the river. Landless households in Saundara thus had relatively greater access to uncultivated land available for common uses. Yet, in the cattle- and buffalo-based dairy development promoted by the cooperative, such goats are not utilized for income. The class dimensions of dairy animal ownership thus limit the participatory potential of cooperative dairying.

Dairying in Mahdol and Saundara is practised in conjunction with crops, and this symbiosis ensures the economic and environmental sustainability of both crop and dairy farming. To the extent that cooperative dairying enables small-scale livelihoods to flourish, it mitigates urban sprawl and rural depopulation. But the link between access to land and cooperative membership shows that the long-term viability of small-scale rural dairying requires either continued land reform or assured access to inputs through the cooperative. If such forms of deliberate inclusion are not implemented, the exclusions of cooperative dairying will match those of the Green Revolution with deleterious results for sustainable forms of both cropping and dairying.

Cattle and buffaloes

The multiple uses of cattle in India prohibit breeding a population of animals exclusively devoted to dairying. We have already considered how the Green and White Revolutions conflict with one another as high-yielding dairy cattle have not been bred for draft uses (George 1990, Nair 1990). Mahdol and Saundara exemplify conditions in which crossbred cows become acceptable and unacceptable, respectively (Basu 2009a).

As mentioned earlier, in the village of Mahdol, crossbred cows are the principal dairy animals and have replaced buffalo. Crossbred cows yield about three times as much milk as buffaloes, and the increased costs of maintenance associated with high-yielding cattle are partly subsidized by the cooperative programme through artificial insemination services, access to a veterinarian, sales of cattle feed and refrigeration for storage of large milk quantities. This dependence on external inputs is problematic to the extent that it makes the village vulnerable to any withdrawal of support by the dairy programme. Zebu cattle are completely absent in the village, except for a few bullocks. The decreased private ownership of zebus is partly due to the small size of land holdings whose ploughing and weeding can be entrusted to hired zebu bullocks at reasonable costs.

In contrast, Saundara's farmers have refused to adopt crossbred cows despite much prodding by the milk union, and buffaloes are the principal dairy animals. This matches the fact that the majority of milk in India is produced by buffaloes (Dairy India Yearbook 2007). The higher fat content of buffalo milk makes it more remunerative than cow's milk since cooperative milk payments are based on a combination of quantity and fat content. More significantly, Saundara's cattle are set to work in the fields. Draft cattle such as zebu bullocks are crucial in monsoons when tractors and other vehicles cannot navigate muddy fields. As one farmer put it:

> Despite the tractor, bullock carts are required. Tractor cannot be sent everyday to bring fodder because diesel is expensive, and tractor does not run in the rains, so bullock cart becomes useful (Interview, Mahdol village, January 2001).

Dairy cows become useful here in ensuring a steady supply of healthy bullocks for draft uses. Their milk thus is more likely to be allotted to feeding calves, rather than being solely used for human consumption. As another farmer expressed it, 'Gai mata [roughly, mother cow] gives us milk, she gives us bullocks, if we didn't own her we would not be able to farm' (Interview, Saundara village, October 2000).

Farmers here enumerated many problems with crossbred cows, all of which devolved around the lack of local knowledge and control over crossbreds. First, given that the Saundara cooperative did not provide veterinary services, the greater amount of health problems associated with crossbreds was daunting, while local cattle and buffaloes could be treated in the village. Second, crossbreds cannot be grazed and fodder must be brought to them, and this extra work was too burdensome for agricultural labourers under conditions of intensified agricultural production. Third, crossbred cattle could not be bought and sold in local markets, making it difficult to sell unproductive animals.

In both villages, dung was extolled as the most useful product from cattle, and one of the advantages of cooperative dairying was that it enabled more animals to be maintained in villages. The value attached to manure indicates not only the symbiosis between crop and dairy economies, but it also shows that the shift to chemical agriculture remains incomplete since both villages utilize organic fertilizers to renew soils. The negative environmental effects of Green Revolution chemical inputs (Shiva 1991, Singh 2000) are thus mitigated by cattle dung, and sale of excess manure is another way cattle contribute to diversifying rural livelihoods.

A member of the household that supplies the largest quantity of milk to the dairy cooperative expressed the link between dairy and access to dung as follows:

> The advantage of buffaloes is we have more khad [fertilizer] at home. [Animal] khad increases agricultural production [specifically, doubles production]. If there was no dairy we would suffer a loss, that is we would not be able to look after so many animals, since people do not have enough arrangements to keep non-productive animals [since there is less land and hence not enough fodder]. The entire village is helped by dairy because of khad. Poor people can sell khad. ... The main benefit is this that we get access to khad (Interview, Saundara village, January 2001).

Another farmer mentioned that 'land gets spoiled due to chemical fertilizers, so we have to use more gobar [cattle dung]' (Interview, Saundara village, January 2001).

The promotion of specialized cattle for dairying imposes a single meaning of productivity onto a landscape in which cattle are also important for crops. The resources to maintain zebus are available locally, while crossbreds promote dependence on external inputs. The adoption, or not, of crossbreds thus should be seen as part of local approaches to dairying and cropping, instead of being judged only by development-led definitions of the need for higher milk yields. However, the probability must be noted that differences in dairy animals between Mahdol and Saundara will lessen as the declining profitability of Green Revolution agriculture leads small farmers towards high-yielding cattle.[5] This is because with the decline in agriculture, dependence on cooperative dairying increases, which facilitates an acceptance of cooperative cows. This is what has happened in Mahdol, and is now happening in Saundara, and it raises questions about the extent to which intensified commodification of farm inputs interferes with symbiotic crop–livestock systems, risking the socio-economic sustainability of small farmers.

Availability of labour

The work of crop and dairy farming was distinctly gendered in both villages. In Mahdol, women from landowning households do not participate in cropping, but consider looking after dairy animals and carrying milk to the cooperative as their principal responsibilities. Since crossbred cows cannot be grazed, women cut and carry fodder from fields to cattle sheds. The decline of agricultural income due to decreasing size of land holdings and increasing costs of agricultural inputs has made cooperative dairying a valuable source of income. As one woman mentioned,

> People [of the village] do not get jobs, because they are not educated. Even educated people do not get jobs. Even if they get jobs they work more and are paid very little. So when women keep animals, dairy helps them take care of the household expenses, schooling of children (Interview, Mahdol village, April 2001).

However, as agriculture continues to decline, this has led men to share the work of dairying with women, with possible implications for women's access to dairy work in the future. This dependence on dairying is more acute in households where men have not had success in migrating to new job opportunities.

In Saundara, women from landed households supervise labourers and themselves work crops in the fields. Few women consider dairying their responsibility, so most dairy activities are in the domain of men's work. Further, hired agricultural labour (male) is often entrusted with cattle-related tasks. The intensification of women's labour here has thus worked in the interests of the Green Revolution rather than the White. This points to the value of understanding regionally specific constructions of women's availability for crop and dairy farming, instead of a priori assuming that women's participation is always available for dairying work.

Sharing of milk in villages can also be linked to forms of labour used for dairying. In Mahdol, where dairy work depends on family labour, milk sufficient for household needs is set aside and the rest is sold to the cooperative. In Saundara, besides household needs, milk is often set aside for agricultural labourers, not merely as payment but also to highlight the social value of being attached to specific farmers. Such forms of sharing were probably more pronounced in the absence of ready markets for milk. In both villages, therefore, landless households report that their ability to buy milk in the village has decreased since the advent of the cooperative. Similar to the Green Revolution, an excessive focus on external markets in the White Revolution thus could leave non-dairy rural households devoid of milk, and further destroy the social and environmental synergies that characterize self-sufficient rural livelihoods.

In both villages, cropping and dairying form an interlinked system, despite greater dependence on cooperative inputs in the case of Mahdol. Thus, agriculture is not a process of specialization but involves the sum total of crops and livestock, suggesting that the Green and White Revolutions cannot be discussed separately in the context of rural India but should be viewed as intersecting entities. Yet, crop–livestock farming has not become a policy goal in rural development. Against this, in order to synchronize local agricultural practices with official rural development policies, there is a need for greater attentiveness to the composite construction of rural livelihoods (Millar and Photakoun 2008, Funes-Monzonte *et al.* 2009).

Conclusion: sustaining crop–livestock systems

Official development agencies generally equate the Green and White Revolutions with increasing yields and, since the 1990s, deepening privatization. However, village-level crop and dairy farmers also prioritize sustainability of social relations and environmental resources. The case studies above show that the economic, social and environmental sustainability of rural livelihoods is linked to drawing on local resources and family labour in both cropping and dairying. This is not to argue that small farmers in India oppose productivity (Nair 1990, Rutten 1995, Gupta 1998,

Ramamurthy 2011), but that productivity is related to sustainability to a greater extent in local contexts than it is in official pronouncements. However, there are unsustainable dimensions to such practices – environmental resources, especially land and water, are unequally distributed, and women's contributions to cropping and dairying are seldom matched by ownership of household assets – and a broader notion of sustainability needs to take these into account.

The recognition of close connections between arable and dairy agriculture in these case studies suggests lessons for the replication of Green and White Revolutions in other contexts. First, there is a need to link productivity to farming practices which combine crops and livestock, including the use of animal manure for the improvement of soil fertility. As Horlings and Marsden (2010, p. 41) note: 'animals are not only a source of manure, but also of farm power and immediate cash. Yet most economic studies leave them out'. Thus, the Green Revolution in Africa is likely to benefit from an increase in livestock numbers if symbiotic relationships between cropping and dairying are encouraged (Otsuka and Kalirajan 2005). New dairy programmes, like the Pro-Poor Livestock Policy Initiative (PPLPI 2012) and the East Africa Dairy Development (EADD 2010) project, thus need to be discussed in conjunction with attempts to build A Green Revolution for Africa (AGRA 2009). Second, the fate of small and landless rural residents continues to depend on combining land and animal resources to build economic self-sufficiency and sustainable soils, considering input price rises in this era of 'peak oil' and prospect of 'peak food' (Horlings and Marsden 2010, pp. 4, 5). Cooperatives thus remain significant for pooling of members' resources, especially in the absence of political will to build and pursue land reform programmes. Third, the labour needs of new farm technologies and their complex repercussions on local social relations should be understood, requiring a cautious engagement with women's work, so that women are not unduly burdened with responsibility for rural development.

The political organizing required to produce successful development also should be kept in mind. The principal lesson of the cooperative dairying-based White Revolution is that political organizing is not antithetical to development, but is an important condition for locally appropriate development. What is meant by successful dairy development depends upon whether it prioritizes milk yield and net profit for a few specialized producers, or benefits as many participants as possible on a range of development indicators, in what can be described as much a social movement as an agricultural development programme. Overall, land and cattle, and labour and technology, are meaningless without a system of mutually beneficial social relationships organized among them, and the White Revolution epitomizes this through people's participation. There are useful lessons here for ensuring the sustainability of rural livelihoods by utilizing the synergy between crop and livestock systems already present in local contexts.

Notes

1. In 1990–1991, 59 per cent of India's farmers owned less than 1 hectare of land and 19 per cent owned between 1 and 2 hectares. See Basu (2009c).
2. This section draws on research conducted by Basu and further details on methods and findings can be found in Basu (2009b). Fieldwork for this research was conducted in 2000–2001 with funding provided by the International Dissertation Research Fellowship (IDRF) Program of the Social Science Research Council (SSRC) with funds from the Andrew W. Mellon Foundation and a Doctoral Dissertation Improvement (DDRI) Award (BCS-0000280) from the National Science Foundation (NSF).
3. Saundara is located along the Narmada river and is one of the villages slated to be submerged by the Maheshwar dam. However, at this time (2012), the village is yet to be submerged, and at the time that the fieldwork for this article was conducted (2000–2001), cooperative dairying in the village continued without reference to impending displacement, especially since the village at that point was successfully resisting the Maheshwar project.
4. Contrasts between the two villages can also be viewed through the condition of roads (a significant aspect of ease of milk transportation) and the extent of dependence on migration. The states of

Gujarat and Madhya Pradesh differed markedly in terms of road quality at the time of the fieldwork. Road quality was not an issue in Mahdol due to the better network and quality of roads in the state of Gujarat. In Saundara, the presence of a paved road, which connected the village to the highway, was an anomaly in the local context and displayed the ability of the village to draw development funding towards itself through its political connections. In terms of migration, Mahdol is situated in a region that has very high rates of international migration and itself seeks to participate in migrant economies. The village however has had less success in comparison to larger surrounding villages. Dairying income becomes important here as provider of subsistence income to support households in the absence of men who have migrated, and as a source of initial capital to facilitate migration. There was no appreciable outmigration from the village of Saundara at the time of fieldwork.

5. Birkenholtz (2009) has shown that farmers in Rajasthan have moved away from high-yielding Green Revolution to traditional crop varieties in the face of agricultural decline. The present paper builds on ethnographic data to suggest a shift from high-yielding crops to high-yielding cattle, but cannot make more generalized conclusions beyond this.

References

Agarwal, B., 1984. Rural women and high yielding variety rice technology. *Economic and political weekly*, 19, A39–A52.

AGRA (Alliance for a Green Revolution in Africa), 2009. *Home*. Available from: http://www.agra-alliance.org/ [Accessed 20 February 2012].

Amul, 2010. *Achievements of dairy cooperatives*. Available from: http://www.amul.com/achievements dairycoop.html [Accessed 20 February 2012].

Basu, P., 2009a. Success and failure of crossbred cows in India: a place-based approach to rural development. *Annals of the association of American geographers*, 99, 746–766.

Basu, P., 2009b. *Villages, women, and the success of dairy cooperatives in India: making place for rural development*. Amherst, NY: Cambria Press.

Basu, P., 2009c. Viewing global futures through rural transformations: lessons from India. *Whitehead journal of diplomacy and international relations*, 10, 83–97, Available from http://blogs.shu.edu/diplomacy/files/archives/07%20PBasu_Layout%201.pdf [Accessed 20 February 2012].

Basu, P. and Chakraborty, J., 2008. Land, labor, and rural development: analyzing participation in India's village dairy cooperatives. *Professional geographer*, 60, 299–313.

Birkenholtz, T., 2009. Irrigated landscapes, produced scarcity, and adaptive social institutions in Rajasthan, India. *Annals of the association of American geographers*, 99, 118–137.

Byres, T., 1981. The new technology, class formation and class action in the Indian countryside. *Journal of peasant studies*, 8, 405–454.

Candler, W. and Kumar, N., 1998. *India: the dairy revolution: the impact of dairy development in India and the World Bank's contribution*. Washington, DC: World Bank.

Chowdhury, P., 1992. *The veiled women: shifting gender equations in Rural Haryana, 1880–1990*. Delhi: Oxford University Press.

Cornwall, A., 2002. *Making spaces, changing places: situating participation in development*. Sussex, UK: Institute of Development Studies.

Cowen, M. and Shenton, R., 1996. *Doctrines of development*. New York: Routledge.

Dairy India Yearbook, 2007. *Dairy India 2007*. Delhi, India: Dairy India Yearbook.

Das, R., 2002. The green revolution and poverty: a theoretical and empirical examination of the relation between technology and society. *Geoforum*, 33, 55–72.

Devendra, C., 2002. Crop-animal systems in Asia: future perspectives. *Agricultural systems*, 71, 179–186.

Devendra, C. and Thomas, D., 2002. Crop–animal systems in Asia: importance of livestock and characterisation of agro-ecological zones. *Agricultural systems*, 71, 5–15.

Dhanagre, D., 1987. Green Revolution and social inequalities in rural India. *Economic and political weekly*, 22, AN137–AN139+AN141–AN144.

Doornbos, M., *et al.*, 1990. *Dairy aid and development: India's operation flood*. Delhi; Newbury Park, CA; London: Sage.

Dubash, N., 2002. *Tubewell capitalism: groundwater development and agrarian change in Gujarat*. Oxford: University Press.

EADD (East Africa Dairy Development), 2010. *About us*. Available from: http://eadairy.wordpress.com/about/ [Accessed 20 February 2012].

Escobar, A., 1995. *Encountering development: the making and unmaking of the Third World*. Princeton, NJ: Princeton University Press.

Finnemore, M., 1997. Redefining development at the World Bank. *In*: F. Cooper and R. Packard, eds. *International development and the social sciences: essays on the history and politics of knowledge*. Berkeley: University of California Press, 203–227.

Foresight, 2011. *The future of food and farming*. London: Government Office for Science.

Funes-Monzonte, F., *et al.*, 2009. Conversion of specialised dairy farming systems into sustainable mixed farming systems in Cuba. *Environment, development and sustainability*, 11, 765–783.

George, S., 1985. *Operation flood: an appraisal of current Indian dairy policy*. Delhi: Oxford University Press.

George, S., 1990. Agropastoral equations in India: intensification and change of mixed farming systems. *In*: J. Galaty and D. Johnson, eds. *The world of pastoralism: herding systems in comparative perspective*. New York: Guilford Press, 119–143. and London: Bellhaven Press.

Gupta, A., 1998. *Postcolonial developments: agriculture in the making of modern India*. Durham, NC: Duke University Press.

Hazell, P., 2011. *Five big questions about five hundred million small farms*. Paper presented at the *IFAD conference on new directions for smallholder agriculture*, 24–25 January 2011. Available from: http://www.ifad.org/events/agriculture/doc/papers/hazell.pdf [Accessed 17 February 2012].

Herrero, M. and Thornton, P., 2010. Mixed crop livestock systems in the developing world: present and future. *Advances in animal biosciences*, 1, 481–482.

Herring, R., 2007. The genomics revolution and development studies: science, poverty and politics. *Journal of development studies*, 43, 1–30.

Hickey, S. and Mohan, G. eds., 2004. *Participation: from tyranny to transformation? Exploring new approaches to participation in development*. London, New York: Zed Books.

Hiremath, B., Singh, K., and Mergos, G., 1997. *Impact of operation flood in Madhya Pradesh: a resurvey*. Anand, India: Institute of Rural Management, Anand (IRMA).

Horlings, I. and Marsden, T., 2010. *Towards the real green revolution? Exploring the conceptual dimensions of a new ecological modernisation of agriculture that could 'feed the world'*. Working Paper Series No. 54. Cardiff: Cardiff University Centre for Business Relationships Accountability, Sustainability and Society (BRASS).

Jayne, T., Mather, D., and Mghenyi, E., 2010. Principal challenges confronting smallholder agriculture in Sub-Saharan Africa. *World development*, 38, 1384–1398.

Katakam, A., 2002. New economic policies, the real threat. *Frontline*. 19. Available from http://www.hinduonnet.com/fline/fl1907/19070900.htm [Accessed 20 February 2012].

Kurien, V., 1997. The AMUL dairy cooperatives: putting the means of development into the hands of small producers in India. *In*: A. Krishna, N. Uphoff and M. Esman, eds. *Reasons for hope: instructive experiments in rural development*. West Hartford, CT: Kumarian Press, 105–119.

Kurosaki, T., 2006. Milk, fodder, and the Green Revolution: the case of mixed farming in the Pakistan Punjab. *Pakistan development review*, 35, 537–548.

Mergos, G. and Slade, R., 1987. *Dairy development milk cooperatives: the effects of a dairy project in India*. World Bank Discussion Paper 15. Washington, DC: World Bank.

Millar, J. and Photakoun, V., 2008. Livestock development and poverty alleviation: revolution or evolution for upland livelihoods in Lao PDR? *International journal of agricultural sustainability*, 6, 89–102.

Moore, J., 2008. Ecological crises and the agrarian question in world-historical perspective. *Monthly review*, 60, 54–62.

Moritz, M., 2010. Crop–livestock interactions in agricultural and pastoral systems in West Africa. *Agriculture and human values*, 27, 119–128.

Nair, K., 1980. Operation Flood: some constraints and implications. *Economic and political weekly*, 15, 446–448.

Nair, K., 1985. White Revolution in India: facts and issues. *Economic and political weekly*, 20, A89–A95.

Nair, K., 1990. Cattle development in Kerala: trends and prospects. *Economic and political weekly*, 25, 2005–2007+2009–2010.

Otsuka, K. and Kalirajan, K., 2005. An exploration of a Green Revolution in Sub-Saharan Africa. *Journal of agricultural and development economics*, 2, 1–6.

Oza, R., 2006. *The making of Neoliberal India: nationalism, gender, and the paradoxes of globalization*. New York and London: Routledge.

Parayil, G., 1992. The Green Revolution in India: a case study of technological change. *Technology and culture*, 33, 737–756.

Patel, S., 1990. The Anand pattern: a socio-historical analysis of its origin and growth. *In*: M. Doornbos and K. Nair, eds. *Resources, institutions and strategies: Operation Flood and Indian dairying*. Delhi; Newbury Park, CA; London: Sage, 27–56.

Phaniraja, K. and Panchasara, H., 2009. Indian draught animals power. *Veterinary world*, 2, 404–407.

PPLPI (Pro-Poor Livestock Policy Initiative), 2012. *About the project*. Available from: http://www.fao.org/ag/againfo/programmes/en/pplpi/about.html [Accessed 20 February 2012].

Rajapurohit, A., 1979. Cross-breeding of Indian cattle: an evaluation. *Economic and political weekly*, 14, A9–A11+A13–A15+A17–A19+A22–A24.

Ramamurthy, P., 1997. Rural women and irrigation: patriarchy, class and the modernizing state in south India. *In*: C. Sachs, ed. *Women working in the environment*. Washington, DC: Taylor and Francis, 103–126.

Ramamurthy, P., 2011. Rearticulating caste: the global cottonseed commodity chain and the paradox of smallholder capitalism in south India. *Environment and planning A*, 43, 1035–1056.

Rosenzweig, M. and Wolpin, K., 1993. Credit market constraints, consumption smoothing, and the accumulation of durable production assets in low-income countries: investments in bullocks in India. *Journal of political economy*, 101, 223–244.

Rutten, M., 1995. *Farms and factories: social profile of large farmers and rural industrialists in West India*. Delhi: Oxford University Press.

Scholten, B., 2010. *India's White Revolution: operation flood, food aid and development*. London: I. B. Tauris, and Delhi: Viva Books.

Scholten, B. and Basu, P., 2009. White counter-revolution? India's dairy cooperatives in a neoliberal era. *Human geography*, 2, 17–28.

Shiva, V., 1988. *Staying alive: women, ecology and development in India*. Delhi: Kali for Women.

Shiva, V., 1991. *The violence of the Green Revolution: third world agriculture, ecology and politics*. London; New York: Zed and Penang: Third World Network.

Singh, R., 2000. Environmental consequences of agricultural development: a case study from the Green Revolution state of Haryana, India. *Agriculture, ecosystems and environment*, 82, 97–103.

Singh, S., 1979. Operation flood II: some constraints and implications. *Economic and political weekly*, 14, 1765 1774.

Srivastava, N. and Srivastava, R., 2010. Women, work, and employment outcomes in rural India. *Economic and political weekly*, 45, 49–63.

Stringer, L., Twyman, C., and Gibbs, L., 2008. Learning from the south: common challenges and solutions for small-scale farming. *Geographical journal*, 174, 235–250.

The Royal Society, 2009. *Reaping the benefits: science and the sustainable intensification of global agriculture*. London: Science Policy.

Toenniessen, G., Adesina, A., and Devries, J., 2008. Building an alliance for a Green Revolution in Africa. *Annals of the New York academy of sciences*, 1136, 233–242.

Viswanathan, P., *et al.*, 2012. Agrarian transition and emerging challenges in Asian agriculture: a critical assessment. *Economic and political weekly*, 47, 41–50.

Wiggins, S., Kirsten, J., and Llambi, L., 2010. The future of small farms. *World development*, 38, 1341–1348.

World Bank, 2007. *World development report 2008: agriculture for development*. Washington, DC: World Bank.

World Bank, 2012a. *South-South initiative launched to share development knowledge*. Available from: http://web.worldbank.org/WBSITE/EXTERNAL/NEWS/0,contentMDK:21936982~pagePK:34370~piPK:34424~theSitePK:4607,00.html [Accessed 20 February 2012].

World Bank, 2012b. *Can India's dairy revolution be repeated in Africa?* Available from: http://web.worldbank.org/WBSITE/EXTERNAL/NEWS/0,contentMDK:21936907~pagePK:64257043~piPK:437376~theSitePK:4607,00.html [Accessed 20 February 2012].

Yapa, L., 1993. What are improved seeds? An epistemology of the Green Revolution. *Economic geography*, 69, 254–273.

Growing inequality: agricultural revolutions and the political ecology of rural development

Lucy Jarosz

Department of Geography, University of Washington, Seattle, Washington, USA

This article argues that the case studies to be published in a special issue together with this article make three arguments about the deployment of Green Revolution technologies in development strategies. These arguments identify the importance of context in understanding sustainable outcomes, and the claim that technical fixes such as the increasing use of synthetic fertilizers and increased yields do not necessarily alleviate hunger and poverty over the long term. The third argument is that interventions built upon existing social, political and economic inequalities will only deepen unequal access, distribution and control of food and food-producing resources for rural people. In the second half of this essay, these arguments frame responses to the world food crisis of 2007–2008 as exemplified in the Alliance for a Green Revolution in Africa and the International Assessment of Agricultural Knowledge, Science and Technology for Development report. These responses are overshadowed by the large-scale foreign land acquisitions currently underway in many countries where hunger and malnutrition are pressing social and moral issues. Working towards sustainable futures in food and agriculture involves the construction and development of food networks that are contextualized, reliant on agro-ecological paradigms, the provision of dignified and decent livelihoods for food producers, democratically controlled provisioning for surrounding communities and regions as well as secure land rights for small-scale farmers. This is the vision of political ecology of rural development that moves beyond technical fixes and market regulation to considerations of justice and equity in resource access and control.

The case studies from Africa, South Asia and Central America in this volume make three important arguments about the trajectories of rural development that deploy Green Revolution technologies. The first is that understanding the contextualized micro and regional geographies of agriculture and food consumption is vital to building socially and ecologically sustainable futures. Second, rural development interventions that privilege increasing yields and technical solutions have short-term benefits but do not necessarily alleviate hunger and poverty in the long run. And finally, these interventions build on existing inequalities in resource access and distribution over the long term and deepen unequal access, distribution and control of food and food producing resources for farmers, pastoralists and fishers.

I use these three arguments to frame recent responses to rural development trajectories and agricultural development strategies in the wake of the world food crisis of 2007–2008. In the second half of this paper, I examine two responses to this crisis as exemplified in the Alliance for a Green Revolution in Africa (AGRA) and the International Assessment of Agricultural Knowledge, Science and Technology for Development report (IAASTD). Both AGRA and IAASTD focus on small farmers, women farmers and ecological sustainability in their recommendations. But the IAASTD report also emphasizes a key argument emerging from this volume, which is the importance of micro-level and regional ecological, social, economic and cultural contexts in the success of any attempts to increase the sustainability of small-scale agriculture. The report does not make green or gene revolution technologies the centrepiece of its rural development strategies. AGRA tends to emphasize the entire continent of Africa and upscaling the technological fixes defining the Green Revolution along with genetic modification and biofortification of key staples such as sweet potatoes and rice.

The increase in nitrogen-based fertilizers has been a significant short-term investment strategy in the wake of the food crisis. Synthetic fertilizer is a significant feature of the Green Revolution toolkit and may well increase yields in the short term for some farmers. But, as the Guatemala and Malawi case studies in this volume show, fertilizers' dependence on fossil fuels for manufacture and for transport, their increasing prices and their decreasing effectiveness over the long term, demonstrate that they are not a sustainable solution to spikes in world food prices.

As large-scale, foreign land acquisitions increase in nations hard hit by rising food prices, the technocentric responses of AGRA and the micro-level focus upon low-input, diversified, small-scale organic agriculture of IAASTD are overshadowed by the large-scale dispossession of small farmers as investors scramble to secure food and biofuels for foreign enterprises and their citizens. The grass roots, transnational food sovereignty movement contests these land grabs and demonstrates that the future direction of small-scale agriculture is a key part of the debates surrounding responses to the entangled crises of world food and energy prices and global climate change.

Building sustainable agricultural futures

The goal of building sustainable agricultural futures in socially just and ecologically sustainable ways depends on respecting and taking the views, practices and discourses of farmers and consumers most directly affected by development interventions seriously in both the construction and design of these interventions. This bottom-up approach is sensitive to those who are most vulnerable and powerless, and to context-dependent social and environmental relationships, agricultural and livestock production practices, and specific, contextualized understandings of contemporary environmentalisms, human rights and values in people's struggles for dignity, democracy and freedom from poverty and hunger. Secondly, these studies demonstrate that Green Revolution interventions dedicated to increasing yields, market efficiencies and the elimination of hunger may address food gaps in the short term, but cannot address, and sometimes even deepen, structural inequalities, which produce forms of structural violence such as hunger and poverty. These inequalities are based on resource distribution, access and control of seeds, fertilizer, land, water, livestock and labour and span scales from global to local. And thirdly, in examining the histories of Green Revolutions in Swaziland, India and Guatemala (Terry and Klepek), we find that the social, ecological and production successes and failures build on existing disparities in access to land, labour and other productive resources within local, regional and national political ecologies of food and agricultural development.

In their examination of the intensification of dairying, Basu and Scholten argue for a consideration of context-dependent social relations of production in understanding rural development, drawing evidence from the cooperative model of dairy production, processing and marketing known as India's Anand Pattern. Peter Andersen's chapter on indigenous foods such as the rice-bean in India and Nepal indicates the importance of understanding why farmers and consumers may or may not choose to grow or cultivate food plants that are well suited to local ecological conditions and which may contribute to fulfilling nutritional needs. Taste, ease of harvest and a compatibility with a diverse set of livelihood activities shape local choices.

Alan Terry's evaluation of the Green Revolution in Swaziland reveals that residents and farmers in the region benefitted unevenly, and that the 'trickle down' did not reach those not actively involved in Green Revolution of sugar cane production. He demonstrates how uneven local and regional development occurs under the deployment of Green Revolution technology by building on, and deepening, differentials in access to productive resources, and notes that labour access and shortages due to HIV/AIDS are significant in shaping productivity. Marie Javdani also notes how HIV/AIDS in Malawi contributes to inadequate labour resources and thereby negatively shapes the impacts of fertilizer subsidies and increasing yields there. Although showcased as a success story for Green Revolution technology, Javdani notes that yields are shaped by the size of landholdings, weather and how fertilizer distribution is politicized. Access to this input is limited by social relations, which shape decisions about resource distribution.

In Guatemala, Klepek demonstrates that the Green Revolution exacerbated the structural inequities underlying land access and ownership. Fertilizer use increased yield but perpetuated debt. Increasing hunger and poverty were linked to increasingly skewed land access and ownership patterns rather than crop yields. Moreover, corporate intrusion into biotechnology regulation marks external pressures for the adoption of technologies that increase the profits of transnational giants such as Monsanto who monopolize and privatize knowledge and seeds through licensing germplasm and ownership of national seed companies (Middendorf *et al.* 2000, Weis 2007).

Food crisis, AGRA and IAASTD

In the wake of the global food crisis of 2007, researchers have increasingly claimed that the food crisis, the energy crisis, global climate change and the financial crisis are entangled (Araghi 2009, McMichael 2009, Wilkinson 2009). From 2005 to 2008, the international prices of corn and wheat tripled and rice prices soared by 500 per cent (von Braun 2008). Short-term causes included rising oil prices, bad weather in Australia and China, the rise in subsidized biofuels production and financial speculation on agricultural commodities' futures. This confluence of causes led many journalists as well as United Nations' (UN) officials and national leaders to explain the crisis as a 'perfect storm' or 'silent tsunami' that would end once one or more of the short-term causes ameliorated and food prices declined (*Economist* 2008, Jarosz 2009). Global institutions such as the World Bank and the UN have identified the problem as one of supply and demand in the midst of expanding populations (Spieldoch 2008). In his World Food Day ceremony address in Rome, Food and Agriculture Director-General Jacques Diop stressed the need to invest in agriculture and raise the productivity of small farmers in the Global South as well as increasing forms of urban and periurban agriculture (Diop 2009). The World Bank's annual development report in 2008, titled 'Investing in Agriculture', noted the decade's long decline in investment in agriculture research, infrastructure and market development. On 3 June 2008, at the High-Level World Food Security Conference in Rome, UN Secretary-General Ban Ki-Moon declared that food production must rise by 50 per cent by 2030 in order to meet world food security needs. Recent studies affirm that smallholder agriculture and organic farming can feed the world (Halberg *et al.* 2007, Norstad 2007).

Falling food production in Africa dates back to the 1970s. This fact and the global food crisis of 2007 have led to assertions that the Green Revolution technology has bypassed smallholding African farmers who are largely women. The Rockefeller Foundation, the Bill and Melinda Gates Foundation and the UK Department for International Development funded AGRA dedicated to providing 'rapid, sustainable growth in agricultural productivity in Africa among small farmers'. AGRA's objectives are:

> to ensure that smallholders have what they need to succeed: good seeds and healthy soils; access to markets, information, financing, storage and transport; and policies that provide them with comprehensive support. Through developing Africa's high-potential breadbasket areas, while also boosting farm productivity across more challenging environments, AGRA works to transform smallholder agriculture into a highly productive, efficient, sustainable and competitive system, and do so while protecting the environment (AGRA 2010).

This economistic and technical definition of the problem led to an international response emphasizing increasing applications of synthetic fertilizer, pesticides and herbicides, enhanced soil and water management, plant breeding aimed at increased yields, the development of genetically modified crops, the extension of credit and the development of markets (World Bank 2008). These responses resonate with Green Revolution strategies of the past. Overall, forms of low input and organic agriculture are generally sidelined. But organic food production and localized forms of small-scale food production are among the fastest-growing areas in agriculture today and offer great potential in increasing food availability and market development. Agro-forestry and agro-ecology techniques protect the environment over the long term and are inclusive, because they do not depend on high-priced inputs.

Completed in 2008, the IAASTD report, the result of the work of over 400 interdisciplinary experts from around the world, assesses the future of agricultural knowledge, science and technology on poverty and hunger reduction. A number of its key findings contrast with AGRA's objectives. First, the report claims that benefits from increasing agricultural yields were distributed unevenly due to variations among organizational capacities, sociocultural dynamics and institutional and policy contexts. Increasing yields have meant problems with soil and water degradation, increased global warming and eroded biodiversity (Watson 2008). The IAASTD report (2009) stresses interweaving local, indigenous knowledge and technologies with advances in Western sciences and technologies, employing agro-ecological paradigms stressing organic, low-input forms of agriculture, local participation, equity and poverty alleviation. While acknowledging the possibility of future versions of transgenic food plants using recombinant DNA techniques to improve food security, this report observes that promises of drought- or salt-resistant plants have not been fulfilled and does not see a significant role for genetically modified plants currently. Further, the IAASTD report recognizes that small-scale farmers could significantly improve their yields with current non-GM (genetically modified) crops and inputs with improvements in rural infrastructure and the mitigation of post-harvest crop losses (Watson 2008).

Market futures speculation and hoarding also played a role in the food crisis (Clapp 2009, Robles *et al.* 2009). Food and agricultural resources have become increasingly attractive to investors and governments seeking national food and biofuel security. There is no question that oil prices will rise as exploration becomes even more treacherous in the wake of the British Petroleum spill in the Gulf Coast, and the rise in futures grain trading. This will lead to future food crises where prices spike, but will not retreat to former levels (Kaufman 2010). As declared in the media and on the floor of the Food and Agricultural Organization (FAO) headquarters during the World Food Summit: the era of cheap food is coming to an end (Renton 2008, Bello 2009).

Fertilizers

As food prices rose, so did the price of chemical fertilizers. Fertilizer production and transport prices are based on the costs of fossil fuels. Nitrogen fertilizer has been a key driver of increasing world agricultural productivity. On a global scale, total fertilizer consumption has increased from approximately 31 million tonnes in 1961 to 142 million tonnes in 2002. Consumption rates in Asia have gone from zero in the early 1960s to exceeding those of the industrialized countries (IAASTD 2009, p. 151). Average fertilizer consumption in Africa remains low at under 20kg/ha annually as compared with applications of 140kg/ha annually in Asia, where some argue that fertilizers are overused. Many poor farmers cannot afford to buy them in the face of competing demands for medicines, food and school fees, for example. The provision of fertilizer to poor countries was a key response to the food crisis of 2007 (Mosseau 2010, p. 8). Fertilizer inputs comprised over half of the costs of provisional short-term investments in agriculture in 2008 (Mosseau 2010, p. 29). In industrialized countries, synthetic fertilizer accounts for around 30–50 per cent of fossil fuel consumption in intensive cropping systems (IAASTD 2009, p. 183).

Fossil fuels are essential to Green Revolution technologies. Synthetic fertilizers and pesticides fundamental to modern agriculture and its astounding yields are also major contributors to greenhouse gases such as nitrous oxide. Their costs skyrocket in lockstep with historical rises in world oil prices. Producing and distributing synthetic fertilizers and pesticides requires over 5 gallons of fuel per acre. Forty per cent of the world's grain feeds livestock and poultry as the demand for grain-fed meat consumption rises – particularly in the industrialized nations (Popp *et al.* 2010). The Guatemala case (Klepek) reveals that half of Mayan corn production has depended on applications of synthetic fertilizer since the 1960s, but rising costs and decreasing effectiveness have led to a shift to local composting to meet fertilizer needs. Throughout the 1960s and 1970s, fertilizer consumption increased over 500 per cent and is linked to large-scale cotton monocropping. The use of fertilizers by small farmers was touted as one way to lessen their dependency on seasonal plantation labour. However, because of the links between fossil fuels and the costs of fertilizer, fertilizer use reproduced cycles of debt dependence on plantation labour over time, rather than alleviating it whenever oil prices increased. As Klepek points out, Guatemalan small farmers could buy synthetic fertilizers for $50 per tonne in 1971, but due to the oil crisis in 1973, prices climbed to $225 per tonne and showed a 215 per cent increase by 1975. This resonates with the findings in the Malawi case where state subsidies for fertilizer-increased yields over the short term (Javdani). Taken together, these case studies demonstrate that increased yields through subsidized fertilizer use are possible but not for all farmers, and that over time increasing input prices, declining effectiveness and land hunger erode their social and ecological sustainability.

Responding to the crises: land grabs and the People's Food Sovereignty Forum

As the financial system entered a crisis brought on by lax regulation and risky lending practices in the US, international banks, investment firms and pension funds began to look to food and agricultural resources such as land and water as key areas for increasing investment after the initial food price spike in 2007. Countries, corporations, investment firms and pension funds began entering into land deals to ensure food and biofuel security for national interests and as investment strategies. This rapid escalation of large-scale land deals was the centre of attention at the People's Food Sovereignty Forum, the parallel gathering of hundreds of pastoralists, fisher folk, small-scale farmers, and grass roots anti-poverty and anti-hunger organizations and activists gathered in Rome while the World Food Summit unfolded at the UN's FAO headquarters in October 2009. One of the major concerns at the Forum was the pace and the scale of land acquisitions as countries such as South Korea, Libya, Qatar and Saudi Arabia moved to ensure control of key food producing resources in order to ensure national food security.

India has invested $3 billion in lands in Ethiopia, Kenya, Mozambique, Senegal and Madagascar to ensure national food self-sufficiency (Cherian 2010). Malawi is considering bids for land from China and Djibouti, because of its need for foreign exchange. Emergent Asset

Management is investing in 50,000ha in southern Africa, including Swaziland. This land deal is couched in terms of helping small farmers access seeds and other technologies as well as providing employment. 'And African farmland prices are the lowest in the world', notes their CEO (Hunt 2009).

The Women's Caucus concerns, as voiced at the People's Food Sovereignty Forum, centred on attaining secure land, farm, and home ownership, gaining access to healthcare, education and leadership positions, and addressing displacement from wars and calamities. Caucus members cited lack of control over food and the negative impacts of globalization as their challenges and believed that networks among consumers and farmers needed strengthening, as did women's organizations and coalitions.

It appears that while interest in small farmers is growing and rightly placed, a sole emphasis on technologies which build on inequities and disparities fostered, among other things, by the diffusion of Green Revolution technologies, will be inadequate to counter the rapid, surging pace of large-scale land deals currently sweeping the African continent while serving the food, fuel and investment priorities of the investors who are located in the Americas, the Middle East, the European Union (EU), and in South and East Asia. The majority of land grabs centre on a continent with rich resources on offer at bargain basement prices where the majority of people still live in rural areas and make their living from the land. The struggle for the continued existence of small farmers will intensify.

While previous Green Revolutions have increased the volume of yields over the past 40 years, their benefits have not been equality distributed since they tend to unfold in areas most amenable to irrigated forms of agriculture with high subsidies on fertilizer and pesticide. Resource-poor areas that are ecologically diverse, dependent on rain-fed agriculture and home to most of the world's poor are under-served. Farming systems that improve productivity in these places, while using little or no synthetic inputs need, study and application so that local populations can be fed (IAASTD 2009, p. 540).

The current shortfalls in international funding for rural development in the Global South will continue due to fiscal austerity programmes and growing debt concerns in the EU and US. Efforts to mitigate climate change and restructure agriculture point to a current opportunity. There is a growing demand for organic food, an interest in the emergence and development of urban and periurban agriculture emphasizing low inputs such as organic compost to enhance soil structure, labour intensive forms of agriculture subscribing to nitrogen-fixing legumes and trees, cropping diversities over monocropping and a blending of indigenous, contextualized agricultural knowledge with Western research and development. In Africa, improved agro-forestry practices can substantially increase maize yields on degraded tropical soils (IAASTD 2009, p. 175). The government of the Philippines is promoting the use of organic fertilizers and the system of rice intensification to enable poor farmers to increase their yields while reducing their reliance on chemical fertilizer (Mousseau 2010). These agro-ecological practices must be located within a contextualized poverty alleviation strategy involving land tenure security, educational and health care access and building local organizations and cooperatives to more effectively represent their members, whether they are farmers, fishers, pastoralists, women, consumers or indigenous peoples.

Working towards the realization of sustainable futures in agriculture involves the construction and development of food networks that are contextualized and built from a commitment to food provisioning that is ecologically sustainable and reliant on agro-ecological paradigms stressing diversity, low input and circularity rather than a reliance on patented technologies and technical fixes controlled by transnational corporations. These food networks provide dignified and decent livelihoods for producers and democratically provision surrounding communities and regions with safe, healthy and culturally appropriate foods that are accessible and affordable to all residents. This vision fulfils the promise of the political ecology of rural development by recuperating the control and development of food provisioning systems from a focus on markets, commodities and profits, which have tended to deepen inequalities to food and food producing resources, to building relationships among producers and eaters in particular places that also addresses issues of land and water access to small-scale producers, food sovereignty and ecological sustainability that moves beyond from the temporary fix of Green Revolution technologies.

Released in January 2011, *Global Food and Farming Futures* authored by the UK's Foresight Programme, which gathers representatives of industry, government and academia to consider and make policy recommendations on pressing, contemporary social and environmental issues, recommended that policymakers implement interconnected initiatives to encompass the considerations of global food systems with energy needs, water use and quality, land use, sustainable fisheries, climate change and questions of sustainable food production, processing, transport and consumption (Foresight Programme, 2011). This report points out the need for long-term integrated strategies for sustainable agricultural development that consider food needs as encompassing more than nutrition and food security but also drawing from culture, community practices and local knowledge. The report stresses the importance of considering the integrated and global nature of food systems and argues for a more fair distribution of food and increasing attention to those food systems that are not sustainable in terms of energy consumption, rates of environmental degradation and waste. This report underscores the IAASTD's conclusions that food policy is too important to be considered only in terms of market regulation and Green Revolution technologies. The future of food policy must encompass considerations of social justice, environmental preservation and the eradication of world hunger through more equitable systems of production and distribution, which embed the diverse and geographically specific social ecologies and the cultural meanings and practices of food into holistic, interconnected policies and practices.

Note

This article has been amended since originally published online on 9 November 2011 to reflect the removal of one case study by Bhardwaj which unfortunately could not be included in the finished issue.

References

AGRA (Alliance for a Green Revolution in Africa), 2010. *What is the alliance for a green revolution in Africa?* Available from: www.agra-alliance.org/ [Accessed 23 June 2010].

Araghi, F., 2009. Accumulation by displacement: global enclosures, food crisis, and the ecological contradictions of capitalism. *Review*, 32 (1), 113–146.

Bello, W., 2009. *The food wars*. London and New York: Verso.

Cherian, R., 2010. Grabbing Africa. *Frontline*, 27 (9). Available from: http://farmlandgrab.org/12231 [Accessed 1 July 2010].

Clapp, J., 2009. Responding to food price volatility and vulnerability: considering the global economic context. *In*: J. Clapp and M.J. Cohen, eds. *The global food crisis: governance challenges and opportunities*. Waterloo, Canada: Center for Governance and International Governance Innovation and Wilfred Laurier Press, 43–58.

Diop, J., 2009. *Address by the Governor-General at the World Food Day Ceremony, Rome, 16 October*. Available from: www.fao.org/getinvolved/worldfoodday/standup/en/ [Accessed 1 July 2010].

Economist, 2008. The silent tsunami. *Economist*, April 17.

Foresight Programme, 2011. *Global food and farming futures*. Available from: www.bis.gov.uk/foresight/our-work/projects/current-projects/global-food-and-farming-futures/reports-and-publications [Accessed 15 March 2011].

Halberg, N., Alroe, H.L., Knudsen, M.T., and Kristenson, E.S., 2007. *Global development of organic agriculture*. London: CABI Publishing.

Halweil, B., 1999. *Underfed and overfed: the global epidemic of malnutrition*. Washington, DC: World Watch Paper 150, World Watch Institute.
Hunt, K., 2009. Africa investment sparks land grab fear. *BBC news*, 5 August. Available from: http://farmlandgrab.org/6659 [Accessed 1 July 2010].
IAASTD International Assessment of Agricultural Science, Technology and Development, 2009. *Agriculture at a crossroads: global report*. Washington, DC: Island Press.
Jarosz, L., 2009. The political economy of global governance and the world food crisis: the case of the FAO. *Review*, 332 (1), 37–60.
Kaufman, F., 2010. The food bubble. *Harper's magazine*, July, 27–34.
McMichael, P., 2009. The world food crisis in historical perspective. *Monthly Review*, 61 (3), 32–47.
Middendorf, G., Skladany, M., Ransom, E., and Busch, L., 2000. New agricultural biotechnologies: the struggle for democratic choice. *In*: F. Magdoff, J.B. Foster and F.H. Buttel, eds. *Hungry for profit: the agribusiness threat to farmers, food, and the environment*. New York: Monthly Review Press, 107–124.
Mousseau, F., 2010. *The high food price challenge: a review of responses to combat hunger*. Oakland, CA: The Oakland Institute.
Norstad, A., 2007. *Africa can feed itself*. Oslo: The Development Fund.
Popp, A., Lotze-Campen, H., and Bodirsky, B., 2010. Food consumption, diet shifts and associated non-CO_2 greenhouse gases from agricultural production. *Global environmental change*, 20 (3), 451–462.
Renton, A., 2008. Is this the end of cheap food? *The observer*, January 20 Available from: www.guardian.co.uk/lifeandstyle/2008/jan/20/foodanddrink.fooddrinks [Accessed 1 July 2010].
Robles, M., Torero, M., and von Braun, J., 2009. *When speculation matters*. Washington, DC: International Food Policy Research Institute.
Spieldoch, A., 2008. The food crisis and global institutions. *Foreign policy in focus*, August 5. Available from: www.fpif.org/articles/the_food_crisis_and_global_institutions [Accessed 1 July 2010].
von Braun, J., 2008. The food crisis isn't over. *Nature*, 456, 701.
Watson, R.T., 2008. *Testimony to the Financial Services Committee of the US House of Representatives*, Available from: www.agassessment.org/docs/WatsonTestimony514082.pdf [Accessed 1 July 2010].
Weis, T., 2007. *The global food economy*. London and New York: Zed Books.
Wilkinson, J., 2009. The emerging global biofuels market. *Review*, 32 (1), 91–112.
World Bank, 2008. *More and better investment in agriculture*, Word Development Report. Available from http://econ.worldbank.org/WBSITE/EXTERNAL/EXTDEC/EXTRESEARCH/EXTWDRS/EXTWDR2008/0,contentMDK:21498707~pagePK:64167689~piPK:64167673~theSitePK:2795143,00.html [Accessed 1 July 2010].

Index

Page numbers in **bold** type refer to figures
Page numbers in *italic* type refer to tables

For Product Safety Concerns and Information please contact our EU representative GPSR@taylorandfrancis.com Taylor & Francis Verlag GmbH, Kaufingerstraße 24, 80331 München, Germany

Batch number: 10731059

Printed by Printforce, the Netherlands